Planning and Socioeconomic Applications

Geotechnologies and the Environment

Volume 1

The "Geotechnologies and the Environment" series is intended to provide specialists in the geotechnologies and academics who utilize these technologies, with an opportunity to share novel approaches, present interesting (sometimes counter-intuitive) case studies, and most importantly to situate GIS, remote sensing, GPS, the internet, new technologies, and methodological advances in a real world context. In doing so, the books in the series will be inherently applied and reflect the rich variety of research performed by geographers and allied professionals.

Beyond the applied nature of many of the papers and individual contributions, the series interrogates the dynamic relationship between nature and society. For this reason, many contributors focus on human-environment interactions. The series are not limited to an interpretation of the environment as nature per se. Rather, the series "places" people and social forces in context and thus explore the many socio-spatial environments humans construct for themselves as they settle the landscape. Consequently, contributions will use geotechnologies to examine both urban and rural landscapes.

For further volumes:
http://www.springer.com/series/8088

Jay D. Gatrell · Ryan R. Jensen
Editors

Planning and Socioeconomic Applications

Springer

Editors

Dr. Jay D. Gatrell
School of Graduate Studies and
Department of Geography
Geology, & Anthropology
Terre Haute IN 47809
USA
jgatrell@isugw.indstate.edu

Dr. Ryan R. Jensen
Brigham Young University
Department of Geography
Provo UT 84602
690 SWKT
USA
ryan.jensen@byu.edu

ISBN 978-1-4020-9641-9 e-ISBN 978-1-4020-9642-6

DOI 10.1007/978-1-4020-9642-6

Library of Congress Control Number: 2009920104

Printed on acid-free paper

9 8 7 6 5 4 3 2 1

springer.com

To S, my favorite editor, and
our children F, M, E - JDG

To my family - RRJ

Contents

Contributors

Shawn Banasick Department of Geography, Kent State University, Kent, OH, USA

Michael C. Carroll Center for Regional Development and Department of Economics, Bowling Green State University, Bowling Green, OH, USA

Fredric J. Donaldson Department of Geography, Brigham Young University, Provo, UT, USA

Joseph P. Frizado Department of Geology, Bowling Green State University, Bowling Green, OH, USA

Trevor Fuller Department of Geography, University of Illinois at Urbana-Champaign, Urbana, IL, USA

Jay D. Gatrell School of Graduate Studies and Department of Geography, Geology, & Anthropology Terre Haute, IN, USA

Robert Hanham Department of Geology and Geography, West Virginia University, Morgantown, WV, USA

Nancy Hoalst-Pullen Department of Geography and Anthropology, Kennesaw State University, Kennesaw, GA, USA

Richard J. Hoch Department of Geography and Regional Planning, Indiana University of Pennsylvania, Indiana, PA, USA

Ryan R. Jensen Department of Geography, Brigham Young University, Provo, UT, USA

Daniel P. Johnson Indiana University-Purdue University Indianapolis, Indianapolis, IN, USA

Eric W. LaFary The University of Auckland, School of Geography, Geology & Environmental Science, Auckland, New Zealand

Greg H. Lindsey Hubert H. Humphrey Institute of Public Affairs, University of Minnesota, Minneapolis, MN, USA

Vijay Lulla Department Geography, Geology, & Anthropology, Indiana State University, Terre Haute, IN, USA

Fraser Morgan Landcare Research, Auckland, New Zealand; The University of Auckland, School of Geography, Geology & Environmental Science, Auckland, New Zealand

Samuel M. Otterstrom Department of Geography, Brigham Young University, Provo, UT, USA

James Pace Department of Geography, Geology & Anthropology, Indiana State University, Terre Haute, IN, USA

Mark W. Patterson Department of Geography and Anthropology, Kennesaw State University, Kennesaw, GA, USA

Karinne Rancie Department of Geography, Brigham Young University, Provo, UT, USA

Neil Reid Urban Affairs Center and Department of Geography and Planning, University of Toledo, Toledo, OH, USA

Jeffrey M. Sanders Department of Geography, Brigham Young University, Provo, UT, USA

Madhuri Sharma Department of Geography, The Ohio State University, Columbus, OH, USA

Bruce W. Smith Center for Regional Development and Department of Geography, Bowling Green State University, Bowling Green, OH, USA

J. Scott Spiker Independent Scholar, Solon, OH, USA

Rajiv Thakur Department of Earth Sciences, University of South Alabama, Mobile, AL, USA

Jeffrey S. Wilson Department of Geography, Indiana University-Purdue University Indianapolis, Indianapolis, IN, USA

Chapter 1
Geotechnologies in Place and the Environment

Jay D. Gatrell and Ryan R. Jensen

1.1 About This Series

This book is one of the initial contributions to the "Geotechnologies and the Environment" series. The series is an extension of two previous volumes edited by Jensen, Gatrell and McLean (2007, 2005) and is intended to provide an opportunity for specialists in the geotechnologies and academics who utilize these technologies with an opportunity to share novel approaches, present interesting (sometimes counter-intuitive) case studies, and most importantly to situate GIS, remote sensing, GPS, the internet, new technologies, and methodological advances in a real world context. In doing so, the books in the series will be inherently applied and reflect the rich variety of research performed by geographers and allied professionals.

Beyond the applied nature of many of the papers and individual contributions, the series interrogates the dynamic relationship between nature and society. For this reason, many contributors will focus on human-environment interactions. Having said this, the series—including chapters in this book—are not limited to an interpretation of the environment as nature per se. Rather, the series "places" people and social forces in context and thus will explore the many socio-spatial environments humans construct for themselves as they settle the landscape. Consequently, contributions will use geotechnologies to examine both urban and rural landscapes.

1.2 About This Book

Remote Sensing and GIS technologies have become essential tools for professional planners. Yet, the application of these technologies have been rather limited. That is to say, land use planners and similar professionals have tended to use these technologies to simply more quickly produce better maps. Indeed, a survey of public GIS resources on the internet demonstrates that these technologies—which have

J.D. Gatrell (✉)
School of Graduate Studies and Department of Geography, Geology, & Anthropology, Terre Haute, IN, USA
e-mail: jgatrell@isugw.indstate.edu

J.D. Gatrell, R.R. Jensen (eds.), *Planning and Socioeconomic Applications*,
Geotechnologies and the Environment 1, DOI 10.1007/978-1-4020-9642-6_1,

expanded exponentially with the success of Google Earth and other online mapping services—tend to focus on the mapping (particularly tax mapping) capacity of these technologies. In part this historical trend toward mapping is due to the historical tension that has existed between human geography and the geotechniques. This tension is based on the tendency for the technologies to more "naturally" be adopted by physical geographers (Liverman et al. 1998; Pickles 1995; Gatrell and Jensen 2008).

While better maps linked to data are useful, the real power of geotechnologies is based in their capacity to facilitate analysis—not merely display information. In this collection, the intent of the various chapters is to examine planning applications that use the analytical potential of GIS and remote sensing to shape decision making. Additionally, the chapters will underscore the importance of interrogating human-environment interactions, charting dynamic neighborhood canopy change over time (Wilson and Lindsey), and using GIS to aid and understand in the economic development process (Reid et al.; Banasick) as well as comprehensive planning (Patterson and Hoalst-Pullen). The chapters in this book also demonstrate the importance of utilizing new technologies or methodological approaches, such as hyperspectral sensors (Lulla and Jensen) and geographically weighted regression (Hanham et al.). Additionally, chaptersof Thakur and Sharma examine the practical complications of "doing GIS and remote sensing" in the developing world.

Table 1.1 Summary of substantive chapters in this book

Chapter No.	Author(s)	Subject
2	Reid et al.	GIS and Economic Development
3	Wilson and Lindsey	Neighborhood Change and The Canopy
4	Hanham et al.	Monitoring Urbanization with Geographically Weighted Regression
5	Morgan and LaFary	GIS and Eco-system Management
6	Lulla	Urban Hyper Spectral Remote Sensing
7	Banasick	Uneven Development and Inequality in Japan
8	Fuller and Gatrell	Urban Environmental Risk
9	Thakur and Sharma	Deploying GIS in Peripheral Regions
10	Johnson	GIS and Medical Geography
11	Pace and Gatrell	Alternative Energy Policy and Planning
12	Rancie et al.	Historical GIS and Urban City-Systems & Counties
13	Patterson and Hoalst-Pullen	Comprehensive Planning Applications

In addition to methods, technique, and more traditional applications, the papers in this volume explore new intersections between resource management and growth (Morgan and LaFary), environmental justice (Fuller and Gatrell), and urban health (Johnson).

1.3 Closing Thoughts

As you read the chapters in this book, we—the editors—believe you will see how the use of GIS, remote sensing and other spatial technologies will enable the disciplines and professions to do what Longley (2002) has referred to as "better" geography. Yet, technology alone does not produce better geographies per se. Rather, the academy and professions must focus on the importance of analysis and synthesis to understand the appropriate place of these technologies and their expanding role throughout society. Hopefully, this book—and the series—will provide scholars and practitioners with a survey of not only contemporary work in the field, but also provide insights into remaining questions, methodological debates, and technical complications of employing geo-technologies in the real world.

References

Gatrell, J. and Jensen, R. 2008. Socio-spatial applications of remote sensing in urban environments. Geography Compass 2:728-743.

Jensen, R. J., Gatrell, J., and McLean, D. (eds) 2005. Geo-Spatial Technologies in Urban Environments. Heidelberg, Germany: Springer-Verlag.

Jensen, R., Gatrell, J., and McLean, D. (eds) 2007. Geo-Spatial Technologies in Urban Environments: Policy, Practice, & Pixels, 2nd Edition. Heidelberg, Germany: Springer-Verlag.

Liverman, D., Moran, E., Rindfuss, R., and Stern, P. (eds.) 1998. People and Pixels: Linking Remote Sensing and Social Science. Washington, DC: National Academy Press.

Longley, P. A. 2002. Geographical information systems: will developments in urban re-mote sensing and GIS lead to 'better' urban geography? Progress in Human Geography 26:231–239.

Pickles, J. (ed). 1995. Ground Truth: The Social Implications of Geographic Information Systems. New York: The Guilford Press.

Chapter 2
GIS and Economic Development

Neil Reid, Michael C. Carroll, Bruce W. Smith and Joseph P. Frizado

Abstract Geographic information systems (GIS) are used by a wide variety of practitioners to help them solve a broad range of spatially-based problems. In this chapter, we focus on the ways in which GIS can be of utility to economic development practitioners who are charged with the task of developing local economies. We begin by defining local economic development (LED). We then report on the extent to which GIS is used by economic development professionals. This is followed by examples of the application of GIS in economic development work. Specifically, we examine the use of GIS in five areas – economic impact analysis, spatial policymaking, identifying potential cluster regions, identifying critical social relationships, and web-based GIS. We conclude the chapter with a few summary statements.

Keywords Economic development · Social network analysis (SNA) · Spatial autocorrelation · Economic cluster

2.1 Introduction

Geographic information systems (GIS) are used by a wide variety of practitioners to help them solve a broad range of spatially-based problems. In this chapter we focus on the ways in which GIS can be of utility to economic development practitioners who are charged with the task of developing local economies. We begin by defining local economic development (LED). We then report on the extent to which GIS is used by economic development professionals. This is followed by examples of the application of GIS in economic development work. Specifically, we examine the use of GIS in five areas – economic impact analysis, spatial policymaking, identifying potential cluster regions, identifying critical social relationships, and web-based GIS. We conclude the chapter with a few summary statements.

N. Reid (✉)
Urban Affairs Center and Department of Geography and Planning,
University of Toledo, Toledo, OH, USA
e-mail: nreid@utoledo.edu

J.D. Gatrell, R.R. Jensen (eds.), *Planning and Socioeconomic Applications*,
Geotechnologies and the Environment 1, DOI 10.1007/978-1-4020-9642-6_2,

2.2 Defining Local Economic Development

The International Economic Development Council (IEDC) defines economic development as "a program, group of policies, or activity that seeks to improve the economic well-being and quality of life for a community, by creating and/or retaining jobs that facilitate growth and provide a stable tax base" (International Economic Development Council 2008).

In discussing local economic development (LED) the World Bank notes that its purpose is to "build up the economic capacity of a local area to improve its economic future and the quality of life for all. It is a process by which public, business and nongovernmental sector partners work collectively to create better conditions for economic growth and employment generation" (World Bank 2008).

We are attracted to the above definitions because we believe that they identify the key characteristics of both the process and objectives of economic development. First, economic development is only possible when the employment base of a community is expanded. Second, the process of expanding the employment base can be best achieved if the various stakeholders (public and private) in a community work collaboratively. Third, the ultimate goal of economic development is a recognizable improvement in the quality of life for the individuals living in a community.

The geographic scale of economic development efforts are broad ranging. At one end of the scale spectrum such efforts focus on particular neighborhoods within a city (Spencer and Ong 2004). At the other end, economic development efforts are supranational in scope, spanning a number of different countries, as in the case of European Union regional policy (European Communities 2004). Regardless of the geographic scale of efforts the work of economic development professionals can be greatly facilitated by the use of GIS technologies. In the remainder of this chapter we outline some of the ways in which GIS can be utilized by economic development professionals. Before doing so, however, we outline the extent to which GIS is used within the economic development profession.

2.3 Use of GIS by Economic Development Professionals

The extent to which GIS is used by local economic development professionals is difficult to gauge as usage data are not systematically compiled. One informative source is a 2004 survey of 464 regional development organizations by the NADO (National Association of Development Officials) Research Foundation. The survey had a response rate of 63 percent. Chase and Thompson (2004) reported that 69 percent of the respondents used GIS and GPS. Although not all regional development organizations in the United States were using GIS, many indicated that they planned to do so in the near future. Only 11 percent of the respondents stated that they were not planning to adopt GIS in the near future.

The primary uses of GIS were in transportation planning (81 percent of respondents), land use planning (75 percent), and community and economic development

(69 percent). The survey also reported that many communities apply GIS in one area, such as community and economic development, but not in other potential areas, such as transportation or land use planning. One major barrier to the expansion of GIS was development organizations' inability to recruit and retain qualified technical staff, especially those organizations located in rural areas.

One trend that we believe will foster the use of GIS is a greater reliance on the internet for the marketing of places by local economic development officials. A 2007 report in Economic Development Online cites a number of people in the site selection industry who emphasize the importance of communities making local economic development information available on the Web (Economic Development Online 2007). For example, Bastian (2002, p. 1) stated that "finding relevant information about 'the next best place' is easier than ever for companies and industry consultants alike, thanks to economic development websites. They save site-seekers time, money, and aggravation." Web sites are a relatively inexpensive method of making relevant community information available to site selectors. As will be discussed later, the integration of GIS and web-based technologies now offer communities some options in terms of using GIS in marketing their community.

2.4 Examples of GIS Use by Economic Development Professionals

In this section we illustrate some of the ways in which GIS can be used by economic development professionals. Five examples are discussed – economic impact analysis, spatial policymaking, identifying the spatial footprint of industrial clusters, examining social relations, and Web GIS.

2.4.1 Economic Impact Analysis

Economic impact analysis has a variety of applications for economic development professionals in a community (Blair 1995). For example, it can be used to gauge the local economic affects of a manufacturing plant closing down. It can be used to estimate the monetary and employment impacts of public and private investments and events, as well as assessing the differential impacts of development options. Examples of applied studies include the impacts of various types of tourism (Bernthal and Regan 2004; Gazel 1998), universities (Blackwell et al. 2002; Carroll and Smith 2006), and manufacturing (Edmiston 2004). There are also methodological discussions of impact studies (Loveridge 2004). To date, with a few exceptions such as Sui (1995) and Carroll and Smith (2006), GIS has been underutilized in economic impact analysis.

To illustrate the contribution of GIS in economic impact studies, we will examine the hypothetical example of a public facility located in Toledo, Ohio. The methodology and data requirements of an economic impact study are briefly outlined, followed by a discussion of the role of GIS in such studies.

2.4.1.1 Input/Output Model

Economic impact studies frequently depend on an input\output model to make the primary economic impact forecasts. With input\output methodology, the forward and backward linkages in a regional economy can be measured (Blair 1995). The model measures the total annual economic activity that results from inter- and intra-industry transactions by subdividing the local economy into approximately 500 separate sectors or individual industries. It then uses a sectoring scheme developed by the IMPLAN Group (MIG, inc. 2004). The final input/output table is approximately a 500 by 500 matrix containing all transactions between the individual sectors. It measures the amount of final consumption by the residents of the region, as well as how much each industry exports from the region. The model uses data collected at the county level, which are obtained from the IMPLAN Group and the BEA (U.S. Department of Commerce 2003). To generate estimates for larger areal units, the county data are aggregated.

2.4.1.2 Data

To conduct an impact analysis of an existing facility, one needs an organization's audited financial statements. These data include accounts payable transactions, salaries, number of employees, construction and capital improvements spending, and the like. All these data typically will contain a geographic anchor, such as addresses, zip codes, or perhaps FIPS codes. Thus, the data can be geocoded to a specific address or areal unit. In our hypothetical example, cities and 3-digit zip codes are the areal units employed to map the accounts payable data since small-scale maps using 5-digit zip codes or counties are too detailed for presentational purposes when showing nationwide data. Other critical data are the geographic origin of visitors, customers, or clients of the facility. Such user utilization data may geocoded to specific addresses or aggregated to some areal unit, such as census tracts or counties. In this example, we use counties to map customer origin data.

2.4.1.3 Defining the Impact Region

A critical first step in any economic impact study is the delineation of the study area, or the impact region, to be analyzed. The impact of a single establishment or a whole industry can be estimated for any defined geographic area. However, Gazel (1998, p. 68) observed that one must consider the area's "economic, political, or social relevance" when defining the impact region. In the case of economic impact studies conducted for local governmental units, Goldman et al. (1997) suggests that one must be sensitive to the needs of local and public officials since they will be using the output of the analysis. Pragmatically, the impact region is frequently a political unit for which published statistics are readily available (Goldman et al. 1997). The economic impact studies of state universities illustrate the significance of the issues of political and economic relevance and data availability. In such studies the impact region is usually the state since state taxes are a major source of revenue for universities and the primary audience for the impact analysis are state legislators, who control state budgets. Despite the importance of carefully defining the impact region, very few impact studies analytically address this issue and often its selection is politically driven (Gazel 1998).

GIS can assist in this process, however, through the utilization of techniques such as overlays and buffers. Gazel (1998) suggests the use of a gravity model to capture the effects of distance-decay in expenditures as one moves away from the local area. Distance decay can be modeled by generating a series of buffers around the facility being studied and then computing the percentage of expenditures in each zone. The buffers can be created using straight-line distances or drive times, depending on which is most appropriate.

In the hypothetical case of the Toledo public facility, we assume that the buffering process demonstrated that a very high percentage of expenditures for our Toledo facility occurred in Indiana, Michigan, and Ohio. Therefore it is apparent that those states, or portions thereof, should be included in the impact region. The next step in the delineation process is to factor in the origins of customers, which we assume have been collected by county. An asset of GIS is that one can create overlays using data compiled for differing areal units. Figure 2.1 shows the distribution of expenditures by 3-digit zip codes and customers by county in the three states.

Using the hypothetical data, it is apparent that twelve counties in northwestern Ohio, southeastern Michigan, and western Indiana should constitute the impact

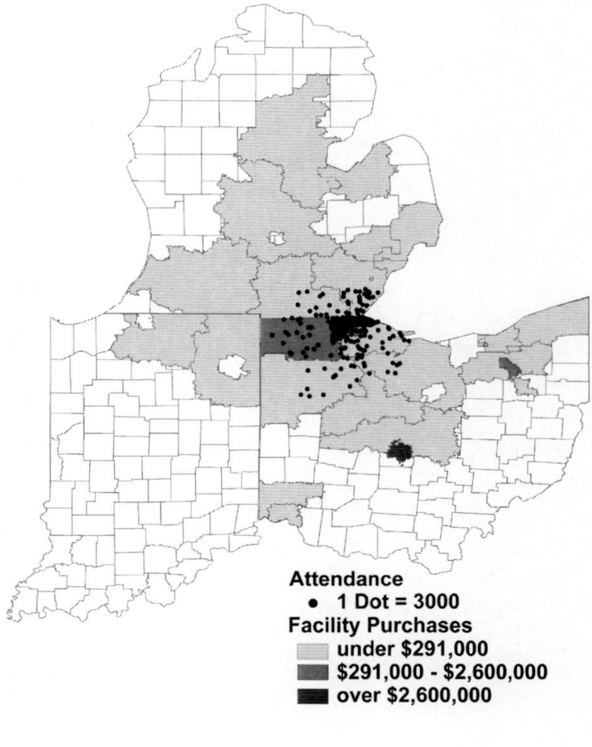

Fig. 2.1 Location of expenditures and customers

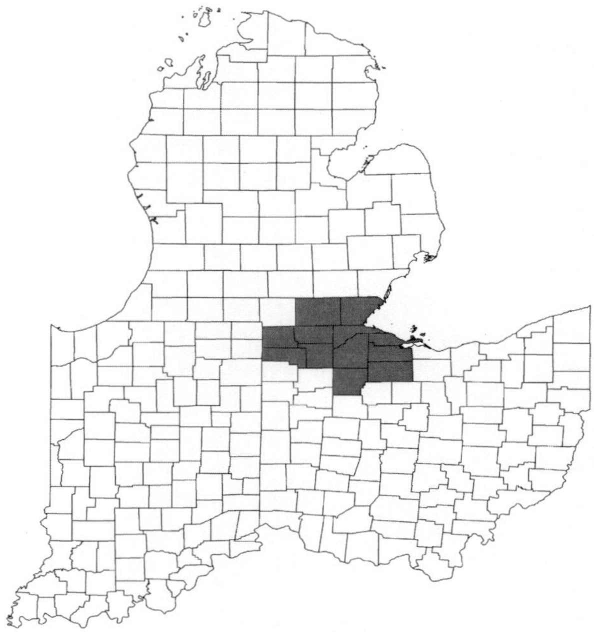

Fig. 2.2 Economic impact region

region (Fig. 2.2). While the final selection of the impact region will always be based on the analyst's judgment, the maps remove some of the subjectivity. Moreover, the justification for defining an impact region in a particular way can be clearly presented to interested stakeholders as a series of maps.

2.4.1.4 Dissemination of Results

A key issue in economic impact studies is the need to effectively communicate findings to a non-technical audience. As Goldman et al. (1997) noted it is important to be sensitive to the interests and questions of the public officials who will use the results of the analysis. With its data visualization and data query capabilities GIS can be very useful in this respect. Non-technical audiences can more readily comprehend the significance of data when it is presented in map rather than tabular form.

Business people are often interested in information about the economic linkages of their facility to different areas of the nation. Use of maps helps convey this information. One can map the distribution of accounts payable by city and show how the organization is linked by their purchasing patterns to various regions of the United States (Fig. 2.3). Thus, the accounts payable transactions can be mapped by city. The ability with GIS to quickly query and map data- sets for alternative areal units is an asset. Frequently people are interested in the economic linkages of their facility at scales other than national. For example, they may want a distribution for an entire state if they are connected to state government. Also they may want the information for various geographic boundaries at the local level. Goldman et al. (1997) suggested

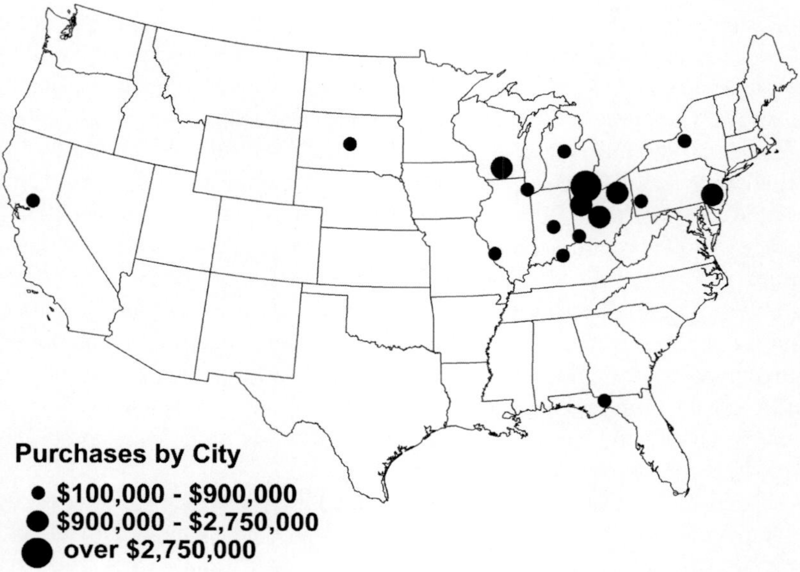

Fig. 2.3 Distribution of accounts payable by city

that if one is examining the impact of various development alternatives within an area then it may initially be necessary to work with overlapping geographic areas until the details of the alternatives are explicitly specified. This is easily accomplished with GIS.

In summary, economic impact analyses have long been used to assess the impact of a variety of developments and events. Even though GIS can contribute significantly to the analysis in an impact study, few studies have incorporated GIS. The advantages of using GIS in an impact study are many. As a database management tool, it facilitates data queries as well as mapping functions for displaying geographical information. The ability to convey data through maps to non-technical audiences is an obvious plus. Also the analytical capabilities of GIS, such as overlays and buffering, are valuable in the identification of the local area of economic impact.

One reason GIS has not been used extensively by economic impact practitioners is the lack of GIS skills. In other cases, data with spatial anchors cannot be acquired. This creates a good opportunity for geographers to collaborate with economists or other economic impact practitioners who have the requisite complementary skills.

2.4.2 Spatial Policymaking

A common application of economic impact analysis is in the creation of a comprehensive economic development strategy (CEDS). A CEDS is required to qualify for Economic Development Administration (EDA) funding (Economic Development Administration 2002). It's purpose is to provide a strategic plan for the project that identifies community participation, a plan of action, and a set of performance measures for project assessment. The economic impact of the project can be placed in the appropriate spatial setting and its impact can be calculated using the I/O model listed above. Generally CEDS are created for small areas such a single county or in rare occurrences multiple counties. Therefore it is useful to break those counties into smaller census units to see actual physical assets in the area. The specific investments that comprise the CEDS are then geocoded and placed within the CEDS area. If there are infrastructure needs associated with a project they are easily identified when the proposed project is integrated with existing city or county utilities. It is certainly easier to alter location of proposed project early in the planning process.

Given the collective nature of a CEDS project, there are considerable discussion and dissemination requirements. It is much easier for policymakers or the general public to see the proposed benefit if the information is present in a map and not long sections of text. Public buy-in can often be completed in a single session if the data are clearly organized and presented in maps.

A second strategic example is the creation of enterprise zones or joint economic development districts (JEDS). These types of projects create an individual political jurisdiction with taxing authority. GIS is useful to identify land owners and business locations before the final spatial footprint is determined for the project. It is often politically beneficial to modify the area if known opposition is to be encountered. This saves a lot of time and can help eliminate prolonged and often messy public debate. GIS also allows a more comprehensive plan to be developed. For example, it is not uncommon for communities to propose the creation of new development parks only to find later that they are located in a flood plain. GIS makes it easy to include topography and hydrology data into the plan.

2.4.3 Identifying Potential Cluster Regions

Another application of GIS is to contribute to the identification of the spatial footprint of industrial clusters. Cluster-based economic development (CBED) has become an increasingly popular economic development tool in recent years. A large number of policymakers and practitioners have adopted CBED as the cornerstone of their region's economic development strategy. For example, the National Governors Association initiated a policy academy in 2007 in which seven states would work on applying cluster analysis and innovation-based economic development strategies in their states (National Governors Association 2007).

While there are numerous alternative definitions of clusters, we believe CBED is most effective when it is viewed as a network-driven economic strategy built upon collaboration among the participants. CBED takes advantage of the positive synergies that can result when cluster members, including business, academia, economic development agencies, and other community members, partner to address the competitive challenges facing a particular industry, which individual businesses, due to lack of resources, cannot successfully address by themselves.

One of the first steps in the process of establishing an industrial cluster is the identification of the cluster's geographical footprint. This can be done by investigating industry location and density patterns to delineate potential cluster regions (PCRs) (Reid et al. 2007). A PCR is a concentration of spatially proximate firms which have the potential to be a cluster. The delineation of PCRs is critical because regions without sufficient concentrations of firms are unlikely to succeed as industrial clusters. It should be noted, however, that even with a sufficient concentration of firms a successful cluster will not develop if industry members, and their non-industry partners, do not engage in joint action.

Various approaches to cluster mapping have been used. In some cases, the spatial boundaries of clusters have been defined politically (Texas Industry Profiles 2004). Others have used key informants to identify clusters (Roberts and Stimson 1998). Another approach has been to use location quotients to identify concentrations of industry, with high location quotients being interpreted as an indicator of a cluster (Miller et al. 2001). More recently other researchers have used spatial autocorrelation measures to delineate cluster boundaries (Feser et al. 2005; Helsel et al. 2006).

2.4.3.1 Methodology

Our suggested approach is to use GIS to identify PCRs. We incorporate both location quotients and Getis-Ord's Gi* in the delineation of PCRs. Location quotients have long been used by analysts as a measure of regional specialization in a particular industry (Blair 1995). In brief, the location quotient compares an area's employment structure with a larger geographic area, such as the state or nation. Generally, areas with location quotients for an industry greater than 1 are considered to be specialized in that industry. For example, if an in industrial sector comprises 20% of an region's total employment, while comprising only 10% of the nation's employment this region would have a location quotient of 2 (20/10%) in that industry.

The second statistic used to identify PCRs is a measure of local spatial autocorrelation, Getis and Ord's Gi*. Gi* identifies "hot spots", or concentrations, in spatial distributions in which areas and their neighbors have similar values of a given phenomena (Mitchell 2005).

The specification of the local neighborhood by the spatial weights matrix is an important element in the computation of local measures of spatial autocorrelation. Alternative definitions of that matrix will lead to differing neighborhoods, and therefore, perhaps the resultant index of spatial autocorrelation. The spatial weights matrix can be defined using rook or queen's measures of adjacency, distance between county centroids, inverse distance function, inverse distance squared, stochastic weights, and the like (Getis and Aldstadt 2004; Mitchell 2005; Wang 2006).

Selection of a spatial weights matrix ideally should be guided by cluster theory, but there is little consensus in the cluster literature as to the appropriate spatial extent of a cluster. Porter (2000) suggested that a cluster can range in its spatial extent size from being concentrated in a single city to being dispersed across a multi-county region. On the other hand, Litzenberger and Sternber (2005) observed there is no precise threshold for clusters. ESRI (2005) suggested that one should experiment with alternative distance functions and then select the one that shows the greatest amount of clustering. Most previous cluster studies using Gi* have used adjacency or first-order nearest neighbors, including the county itself (Feser et al. 2005; Helsel et al. 2006). Alternatively one could select a spatial weights matrix based on the characteristics of the industry being examined. For example, a contiguity spatial weights matrix was used in defining the spatial footprint of the northwest Ohio floriculture cluster because many greenhouse operators see their competition to be located in their home county and neighboring counties (Smith et al. 2007). In the example which follows, an inverse distance matrix was selected for illustrative purposes. The spatial weights matrix is computed by the Spatial Statistics Tool in ESRI's ArcMap 9.2.

2.4.3.2 Example of a Potential Cluster Region

To illustrate the process of identifying PRCs, the example of the Architectural and Engineering Services and Construction (AEC) industry in Indiana, Michigan, and Ohio is used. Data for NAICS 23, Construction, and NAICS 5413, Architectural, Engineering, and Related Services, were compiled from the 2005 County Business Patterns (U.S. Census Bureau 2005). One drawback of this data source is that the Census does not publish data that would reveal information about individual companies; instead the data on employment in some counties for some industries are reported as series of ranges, such as 0 to 19 employees. In those cases, the midpoints of the ranges were used to estimate employment.

Location quotients were computed for employment, using the tri-state total as the benchmark for comparison (Fig. 2.4). A total of 104 counties in the three states have location quotients greater than 1. The counties which are specialized in this industry are scattered widely throughout the three states. It is somewhat surprising that some larger urban counties do not specialize in this industry, such as Cuyahoga County (Cleveland) in Ohio.

Having computed the location quotients for each county, the next step is to compute the Gi* for employment in the three states. As noted previously, an inverse distance function in the Spatial Statistics Tool of ESRI's ArcMap 9.2 was utilized to compute the requisite values (Fig. 2.5). The index for Gi* indicates how similar a county is to its neighbors. A high Gi* value indicates that high valued counties are located near each other and a low Gi* index indicates that low valued counties are near each other (Wang 2006). Moreover a z-test can be used to test the statistical significance of the index, so values greater than 1.65 are statistically significant at the 0.10 level of significance. Figure 2.5 displays the z-scores of the Gi* values for the counties. Those counties with a z-score over 1.65 are located in an arc around the western end of Lake Erie and comprise 20 counties in Michigan, seven counties in Ohio, and none in Indiana.

To obtain the PCRs, the location quotient results and the Getis-Ord results are combined (Fig. 2.6). Those counties with Gi* indices greater than 1.96 and high location quotients (greater than 1.00) are characterized as "AEC PCR", meaning they are specialized in the AEC industry and also are located near other counties

Fig. 2.4 AEC location quotients

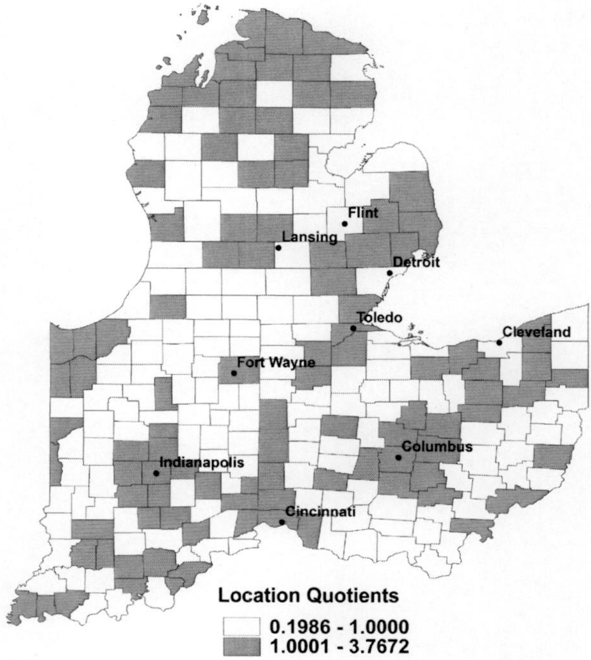

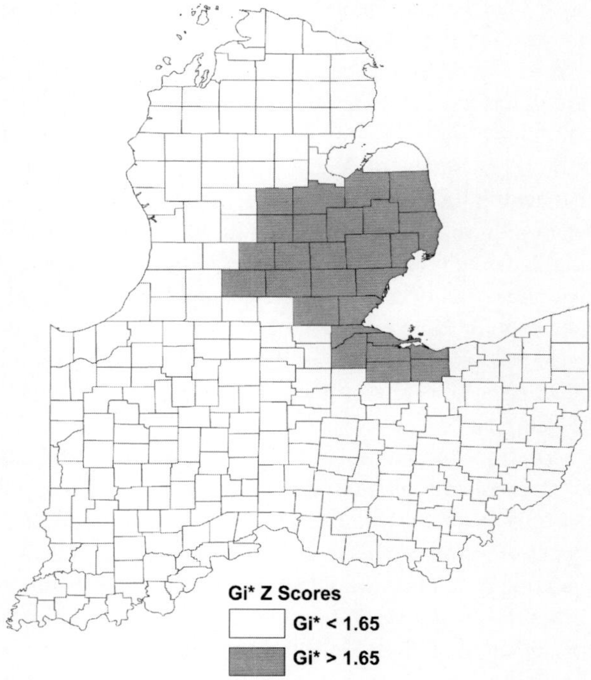

Fig. 2.5 AEC Gi* z-scores

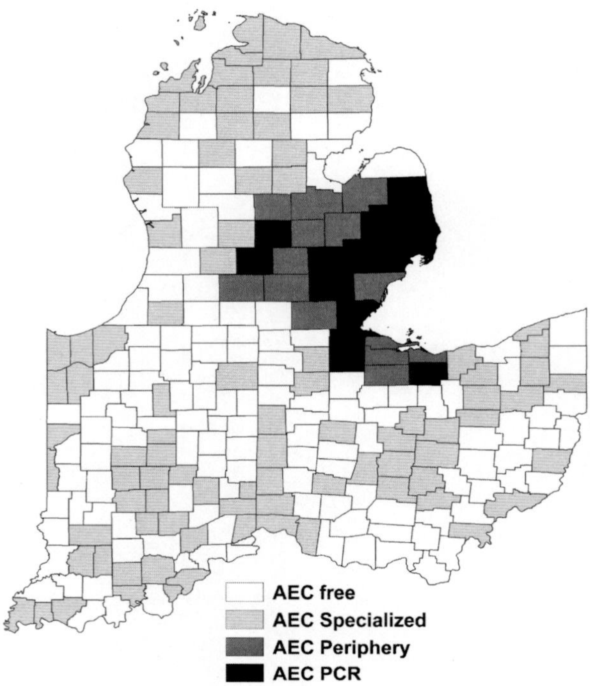

Fig. 2.6 AEC County classification

with high levels of AEC employment. These counties have the highest potential to develop a successful cluster-based economic development strategy focused on the AEC industry. Those counties with high location quotients and low Gi* values are described as "AEC Specialized" because that they contain greater than average amounts of AEC employment. However, they are not candidates for a cluster. While a high percentage of AEC employment is located in them, they are geographically isolated in the sense that their neighboring counties lack sufficient production concentration in this industry for a high Gi*. The "AEC Periphery" counties are those that have high Gi* indices, but low location quotients. Their location near counties which are major centers of AEC employment leads to high Gi* values, but they do not contain large numbers of such employees. Therefore, they lack the industrial base necessary for a viable cluster. The fourth group of counties, the "AEC free", have both low location quotients and Gi* values and have no potential for cluster development.

Both location quotients and measures of spatial autocorrelation have been used to delineate the spatial footprint of a cluster. This approach could be employed with various areal units, such as zip codes, census tracts, and labor markets, assuming the necessary data can be acquired. Both techniques involve some issues. In the case of location quotients, the selection of a particular value of location quotient to use in grouping areal units into a cluster is somewhat arbitrary. In the case of the measures of spatial autocorrelation, the ambiguity derives from the adoption of a particular spatial weights matrix. In particular, the size of the neighborhood is critical. In this example, we used inverse distance, but one could argue that other functions would be equally appropriate.

It should be emphasized that we view the delineation of PCRs as only the first step, in a cluster-based development strategy. Co-location, or geographic proximity, is the primary basis for defining a PCR. A PCR eliminates areas without sufficient production, which will increase the likelihood of the cluster being successful. Nonetheless the sustainability of the cluster is dependent on the willingness of industry members to collaborate with each other and with non-industry partners.

2.4.4 Identifying Critical Social Relations

GIS is beneficial in examining social relationships among actors associated with economic development in an area. Social relationships are a critical component of economic development. The chances that a region will be successful in its economic development efforts will be greatly enhanced when there are healthy relationships between members of the economic development community. Social scientists use the concept of social capital to understand social relationships. Social capital can be approximated as "the stock of active connections among people: the trust, mutual understanding, and shared values and behaviors that bind the members of human networks and communities and make cooperative action possible" (Cohen and Prusak 2001, p. 4). A strong stock of social capital is particularly valuable in a region that is developing a cluster based economic development strategy. Batheldt (2005) proposed that identifying the geography of a cluster should not only consider co-location of businesses, but also their economic interactions and social relations. Similarly, Chetty and Agnadal (2008) view industrial districts as a network of interconnected relationships where actors interact and collaborate in daily activities. In their study of knowledge transfers, Brökel and Binder's (2007) suggested social networks are important because such networks define the spatial range of people's search for information.

In this section the utility of GIS for examining social relations within a cluster or industrial district is illustrated. We examine a hypothetical industry (widgets) which is centered on the city of Toledo, in northwestern Ohio. Further, we assume that there is a cluster of widget manufactures and suppliers in the Toledo metropolitan area (Fig. 2.7). In addition to the manufacturers and suppliers, there are academics, economic development officials, and public officials in the cluster. In this illustration, GIS is combined with social network analysis (SNA), which is a methodology for describing and analyzing social relations. First the basic ideas of SNA are outlined, followed by a discussion of the use of GIS and SNA in examining social relations in an industrial cluster.

2.4.4.1 SNA

SNA is a technique used to describe and analyze the relationships among a group of people and/or organizations (de Nooy et al. 2003). SNA does not consider

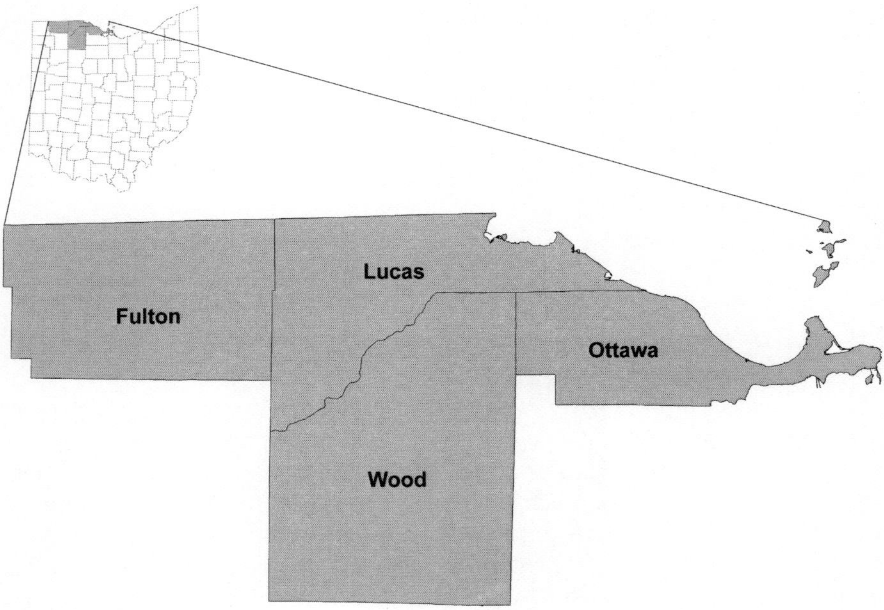

Fig. 2.7 Toledo, OH MSA

individuals as discrete units of analysis; instead it examines the relationships of those individuals. The actions of people and organizations are presumed to be impacted by, and in turn shape, their social networks.

The primary input of SNA is a relational data set describing the links among nodes, which can be individuals or organizations. The relationships between nodes can be kinship ties, business interactions, informational networks, and the like. Figure 2.8 is an illustrative network. It consists of seven nodes, which could be individuals or organizations, and the various links. For example, node A and node E are directly linked. Node E is connected to node G by two steps through node F. Node D is an isolate in the network since it is not connected to any of the other nodes.

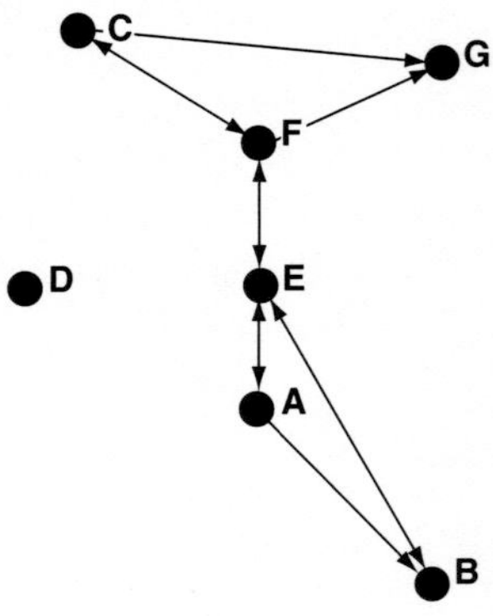

Fig. 2.8 Hypothetical network

Data for a SNA can be collected in various ways (Scott 2000), but the most common method is by use of a questionnaire in which respondents indicate those individuals to whom they are linked. In this example of the widget industry centered in northwestern Ohio, we assume that the data have been collected by asking people in the industry whom they have collaborated with over the past 12 months. Various software programs can be used to analyze social network data (Scott 2000). In this example, UCINET (Borgatti et al. 2002) is used. Most software packages produce an aspatial graphic of the network displaying the nodes and linkages. In addition, metrics describing the network are generated which describe various elements of the network structure, including the nodes and linkages.

The output of a SNA can be very useful to economic developers. For example, it allows the identification of the key players with a region's economic development network and shows who in a network already engage in collaborative relationships. It also allows potential relationship gaps (structural holes) to be identified and bridged. For example, if two people, whom the analysis identifies as being critical, do not have an existing relationship it might be valuable to find a mechanism by which these individuals could develop a relationship. This is referred to as network weaving. Similarly, if the SNA is conducted for an industrial cluster, it shows which individuals are critical to the cluster. Information on the relative importance of individuals within a network can be used to help the network function both more efficiently and effectively. The mapping of individuals and relationships that exist between them can also be informative. As described in the next section mapping the relationships can provide critical insights into the geography of a cluster. Also, in the case of a region's economic development endeavors, mapping relationships can provide insights into the extent of cross-jurisdictional collaboration.

2.4.4.2 GIS Applications

GIS is useful in the examination of industrial clusters. One application is visualization of the geographic distribution of cluster members as well as the total industry. This information is meaningful because not all firms in a given industry will likely chose to participate in a cluster-based economic development initiative, at least initially. For example, Batheldt (2005) noted that the formation of a cluster will not automatically result in the inclusion of all firms which are willing to collaborate on a common set of objectives. Moreover, Watts et al. (2006) found that actor characteristics impacted the extent to which firms are embedded in a cluster. Mapping the evolution of the cluster will provide evidence of its development. Also the maps may provide the basis for analyses of the characteristics of people which impact their willingness to participate in a cluster. Such information may prove insights that will be useful in attracting more businesses to the cluster.

Figure 2.9 is a map showing the distribution of the widget industry in northern Ohio. It displays the locations of both cluster members and non-members. The primary concentration of the industry is in Lucas County, which contains 63.7% of the industry, followed by Wood County with 15.3% of the total. The MSA, which is the cluster region, contains 87.0% of the state's total widget industry. In

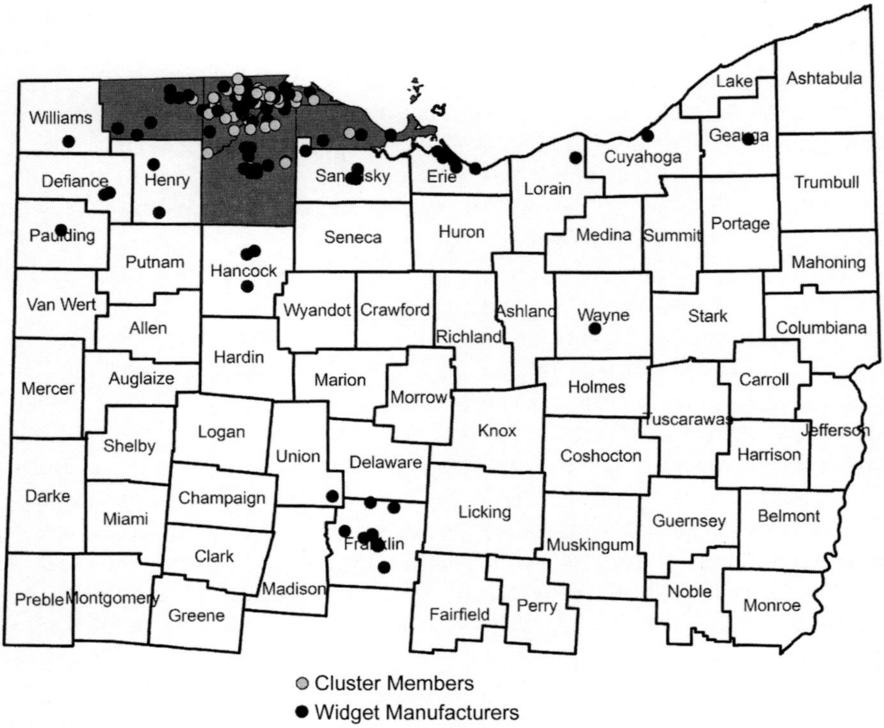

Fig. 2.9 Distribution of widget industry

contrast to the total industry, all of the cluster members are located in the MSA, with 77.8% in Lucas County, 16.7% in Wood County, 3.3% in Fulton, and 2.2% in Ottawa. Figure 2.9 suggests that there is substantial work to be done in expanding the cluster both within the MSA and the larger region of the state. Only 34.4% of the prospective cluster members within the MSA are involved and only 30.0% of the state's total are in the cluster.

The need to expand the cluster is reinforced when one examines collaboration linkages. GIS can be used to display the collaborative linkages among the industry in general and the cluster members (Figs. 2.10 and 2.11). The extent of collaboration among the cluster members is small in comparison to the degree of collaboration among the total industry. Only six percent of the total links are between cluster members. These data suggest that a substantial portion of the collaboration activity in the industry is not included in the cluster. However, 79% of the links occur between a cluster member and a non-cluster member. Those non-cluster members who now collaborate with cluster members would be logical targets for recruitment into the cluster since they are already collaborating with a cluster member.

The geographic distribution of the destinations of the links shows a strong local orientation. Figure 2.12 displays substantial distance decay in the links with over 67% of them being 15 miles or less. Only 7% of the links are over 100

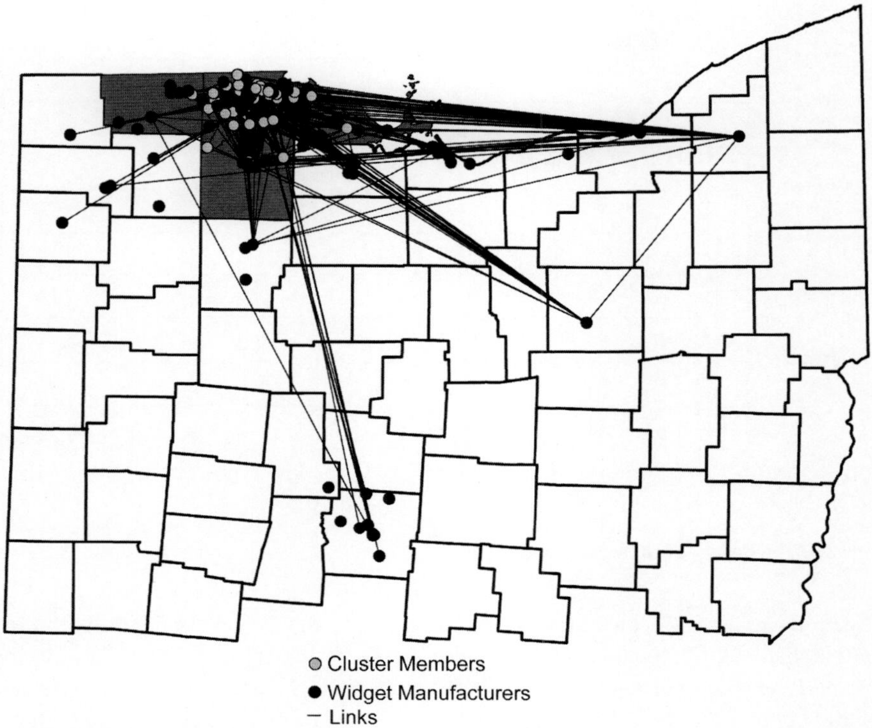

Fig. 2.10 Industry collaborative links

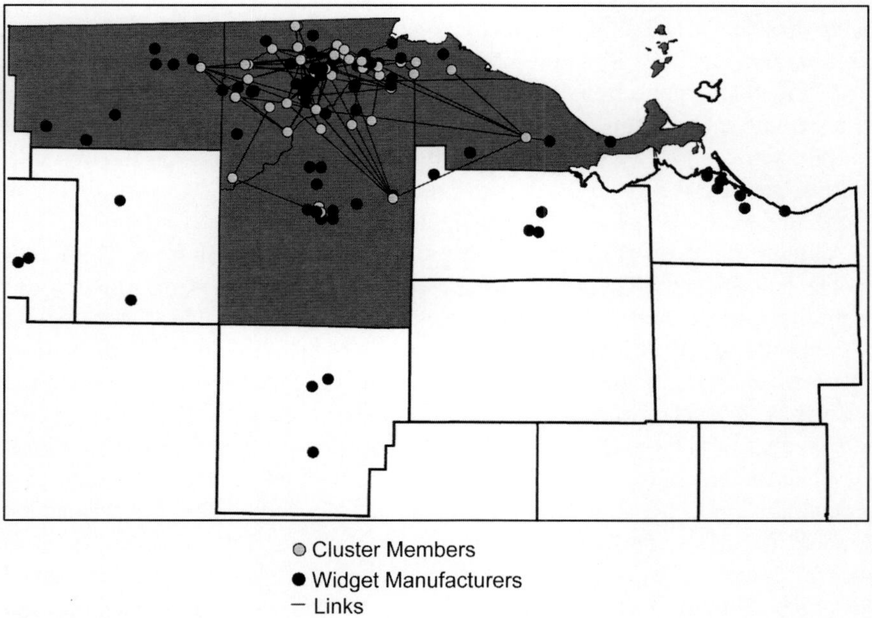

Fig. 2.11 Cluster collaborative links

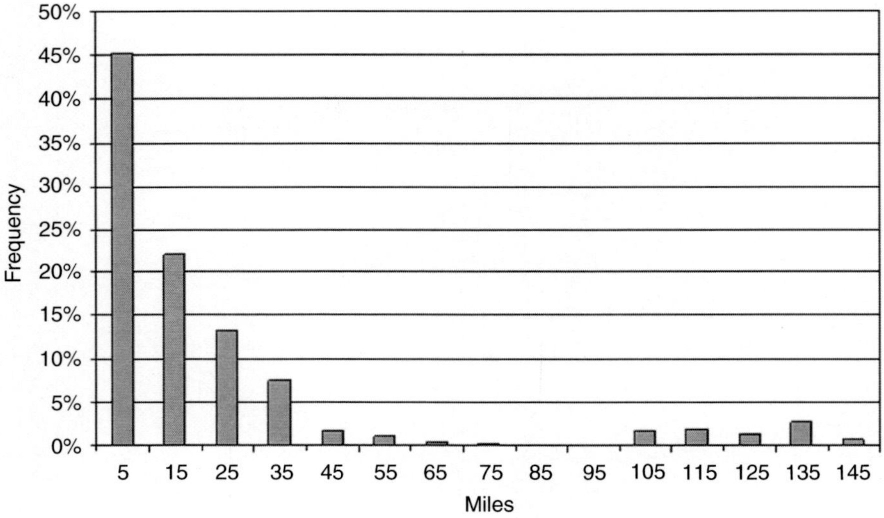

Fig. 2.12 Frequency of linkage distances

miles. In general, the cluster is localized in Lucas and Wood counties. According to Glückler (2007), local ties will be more prevalent than distant connections during the early development of an industry. These local interactions are characterized by Bathelt et al. (2004) as "local buzz", which is the learning, interaction, and collaboration generated by routine activities, such as regular meetings, chance encounters, and social outings.

In addition to local interactions, a network also needs communication channels to the outside world, or "global pipelines" (Bathelt et al. 2004). "Global pipelines" build on Granovetter's ideas on the strength of weak (infrequent) ties (Granovetter 1973). Such connections are sources of new ideas and collaborations as well as providing a fresh outlook in problem solving. The new knowledge will offset the problems of lock-in which can result from too much focus on internal interactions, causing the same ideas to be recycled leading to stagnation or decline (Boschma 2005; Maskell and Malmberg 2007).

Another important attribute of a cluster is the spatial distribution of influential people in the industry. The influence of people in a network can be measured by their centrality. One method of measuring centrality is degrees-in, which is the number of connections that an individual is mentioned by other individuals in the network (Wasserman and Faust 1994). Because degrees-in depends in part on the size of the network, it is standardized by the number of nodes in the network. The normalized value ranges from 0 (no connections) to 1 (connections to everyone in the network).

While the centrality data are aspatial, they can be converted to spatial data by geocoding (Fig. 2.13). There are 52 people with centrality values greater than one standard deviation above the mean and 33 percent of them are a member of the cluster. Moreover, 95 percent of those persons are located in the MSA. A few people with high centrality values are located outside the MSA and the cluster. From the perspective of forming and managing the cluster, it would be desirable to have all

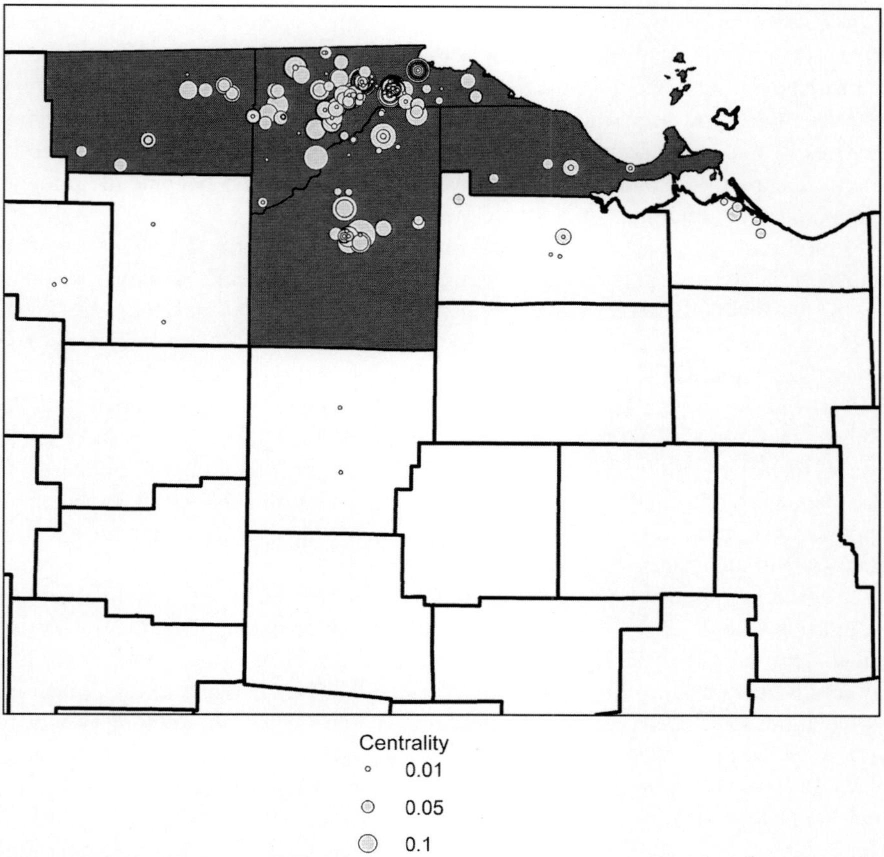

Fig. 2.13 Distribution of centrality scores

people with high centrality, and therefore influence, involved in the cluster, irrespective of their location (Reid et al. 2008). In some cases, clusters are politically defined but collaborative ties among people do not stop at political boundaries. Thus examination of collaborative ties should be one element in establishing the spatial footprint of a potential cluster (Reid et al. 2008).

2.4.5 Web GIS

One practical application of GIS in economic development is using Web GIS to provide data about the local community. As Bastian (2002) noted, economic development web sites are an efficient method by which site selectors can acquire information about potential locations.

The integration of GIS capabilities and web-based technologies has followed two different paths depending upon where the processing is taking place. Full-fledged desktop GIS software can be augmented with server connections to centrally held data centers. Internet connectivity can be transparent to the user where the external data is viewed as 'just another disk' to the desktop. A second approach is to have the server provide all or most of the processing power in response to a request from a browser or lightweight browser-type application (Google Earth, ArcExplorer). Between these end-user situations, applications are being constructed with varying degrees of local or server-processed tasks and datasets. The combination of desktop GIS, server GIS, and browser/lightweight application GIS become the primary components in building an enterprise GIS.

In the past, GIS analysts used desktop systems to perform almost all levels of analysis. Datasets were local and transferred as files via networks or physically sent on media. As interconnection speed has improved and data sets have become larger, network connectivity to large common datasets has become more important. The Open Geospatial Consortium (OGC) has developed protocols from open connectivity to servers while ESRI has also provided a propriety route through ArcGIS Server. Desktop GIS software that provides such connections allows the user to access databases that have been exposed by their owners to public use. Connections are made via URL and the local user can browse the server datasets as if they were on their local computer. Any GIS task can then be performed locally using the external dataset with the results stored on the local computer.

As internet applications have moved into the Web 2.0 realm, one of the most popular is Google Earth. Using a lightweight, browser-like application, a user can view a virtual globe. Using intuitive controls, the user can pan, zoom, and view alternative data layers and information stored on the Google Earth servers. Mash-up applications have been created to overlay the Google Earth representation to portray user-selected information. ESRI (ArcExplorer) and Microsoft (Virtual Earth) have virtual globe applications as well. In each of these examples, most of the GIS processing is done at the server but the boundary between the two is becoming fuzzier. ArcExplorer, when utilized with ArcGIS Server, can allow the server to expose preconfigured tasks that the user can select. For example, any complex geoprocessing analysis can be configured as a task to be run by a distant user on a data set selected by that user. That allows extending web-based mapping beyond finding directions to a location to performing any type of GIS analysis that a geoprocessing server wishes to expose.

Web GIS technology has been utilized to provide data and analytical tools for economic development. Many of these efforts are focused on site selection. The San Francisco Enterprise GIS Program (http://www.sfgov.org/site/gis_index.asp) provides an overview to several online mapping tools for the bay area. SF Prospector (http://www.sfgov.org/site/sfprospector_index.asp) has been providing site selection assistance since 2002. This site provides interactive mapping of demographics, consumer expenditures, vacant commercial properties, and other pertinent data. Users can create demographic analysis and generate site specific business cluster reports based on user-defined selections. Incentive areas, aerial photos, parking, transit, and traffic are viewable online. Massachusetts SiteFinder (http://www.massachusettssitefinder.com/ed.asp?bhcp=1) and ColoradoProspects.com (http://www.coloradoprospects.com/) provide mapping datasets on a more general scale at the state level.

2.5 Summary

This chapter has shown a small sample of GIS applications in economic development. GIS is a powerful tool for data analysis and presentation, and the economic development ramifications are truly significant. The speed at which data and strategies can be coordinated is clearly changing the way economic developers approach their job. There are a number of important trends that are likely to result in GIS becoming more pervasive in the economic development community. These include declining costs of GIS software, increased computing power, and the growth of Web-based GIS applications. There also has been increase in GIS skills among economic development professionals.

References

Bastian, L. (2002). Getting the best from the web. Area Development Site and Facility Planning, March 1–7. Accessed 5 September 2008.

Batheldt, H. (2005). Geographies of production: growth regimes in spatial perspective (II) – knowledge creation and growth in clusters. *Progress in Human Geography*, 29(2), 204–216.

Bathelt, H., Malmberg, A., Maskell, P. (2004). Clusters and knowledge: local buzz, global pipelines and the process of knowledge creation. *Progress in Human Geography*, 28(1), 31–56.

Bernthal, M., Regan, T. (2004). The economic impact of a NASCAR racetrack on a rural community and region. *Sports Marketing Quarterly*, 13(1), 26–34.

Blackwell, M., Cobb, S. Weinbert, D. (2002). The economic impact of educational institutions: Issues and methodology. *Economic Development Quarterly*, 16(1), 88–95.

Blair, J. (1995). *Local Economic Development, Analysis and Practice*. Thousand Oaks, CA: Sage Publications.

Borgatti, S., Everett, M., Freeman, L. (2002). Ucinet for windows: Software for social network analysis. Harvard, MA: Analytic Technologies.

Boschma, R. (2005). Proximity and innovation: A critical assessment. *Regional Studies*, 39(1), 61–74.

Brökel, T., Binder, M. (2007). The regional dimension of knowledge transfers – A behavioral approach. *Industry and Innovation*, 14(2), 151–175.

Carroll, M., Smith, B. (2006). Estimating the economic impact of universities. *The Industrial Geographer*, 3(2), 1–12.

Chase, M., Thompson, L. (2004). *GIS Technology: Enhancing Regional Planning & Development.* Washington D.C.: NADO Research Foundation. http://www.nado.org/pubs/gisreport.pdf. Accessed 21 July 2008.

Chetty, S., Agnadal, H. (2008). Role of inter-organizational networks and interpersonal networks in an industrial district. *Regional Studies*, 42(2), 175–187.

Cohen, D., Prusak, L. (2001). *In good company: How social capital makes organizations work*, Harvard Business School: Boston.

de Nooy, W., Mrvar, A., Batagelj, V. (2005). *Exploratory network analysis with Pajek*. Cambridge: Cambridge University Press.

Economic Development Administration (2002). *Comprehensive economic development strategy: CEDS guidelines*. Washington D.C.

Economic Development Online (2007). Economic development and the internet. http://www.economicdevelopmentonline.com/overview.htm. Accessed 5 September 2008.

Edmiston, K. (2004). The net effects of large plant locations and expansions on county employment. *Journal of Regional Science*, 44(2), 289–319.

ESRI (2005). Spatial statistics for commercial applications. *An ESRI White Paper*. http://www.esri.com/library/whitepapers/pdfs/spatial-stats-comm-apps.pdf. Accessed 5 September 2008.

European Communities (2004). *Working regions*. Luxembourg: Office for Official Publications of the European Communities.

Feser, E., Sweeney, S. H., Renski, H. (2005). A descriptive analysis of discrete U.S. industrial complexes. *Journal of Regional Science*, 45(2), 395–419.

Gazel, R. (1998). The economic impacts of casino gambling at the state and local levels. *Annals of the American Academy of Political and Social Science*, 556(1), 66–84.

Getis, A., Aldstadt, J. (2004). Constructing the spatial weights matrix: Using a local statistic. *Geographical Analysis*, 36(2), 90–104.

Granovetter, M. (1973). The strength of weak ties. *American Journal of Sociology*, 78(6), 1360–1380.

Glückler, J. (2007). Economic geography and the evolution of networks. *Journal of Economic Geography*, 7(5), 619–634.

Goldman, G., Nakaszawa, A., Taylor, D. (1997). Determining economic impacts for a community. *Economic Development Review*, 15(1), 48–51.

Helsel, J., Kim, H., Lee, J. (2006). An evolutional model of U.S. manufacturing and services industries. In J. Gatrell, and N. Reid (Eds.), *Enterprising Worlds: A Geographic Perspective on Economics, Environments & Ethics* (pp. 83–98). Dordrecht, Netherlands: Springer.

International Economic Development Council (2008). Frequently asked questions about IEDC. http://www.iedconline.org/?p=FAQs. Last accessed 5 September 2009.

Litzenberger T., Sternber R. (2005). Regional clusters and entrepreneurial activities: Empirical evidence from German regions. In C. Karlsson, B. Johannson, R. Stough (Eds), *Industrial clusters and inter-firm networks* (pp. 260–302). Northamton, MA: Edward Elgar.

Loveridge, S. (2004). A typology and assessment of multi-sector regional economic impact models. *Regional Studies*, 38(3), 305–317.

Maskell, P., and Malmberg, A. (2007). Myopia, knowledge development and cluster evolution. *Journal of Economic Geography*, 7(5), 603–618.

MIG, Inc. (2004). *IMPLAN user's guide*. Stillwater, MN: Minnesota IMPLAN Group, Inc.

Miller, P., Botham, R., Martin, R., Moore, B. (2001). *Business clusters in the UK: A first assessment*. London: Department of Trade and Industry.

Mitchell, A. (2005). *The ESRI guide to GIS analysis, Volume 2: spatial measurements and statistics*. Redlands, CA: ESRI Press.

National Governors Association. (2007). *Cluster-based strategies for growing state economies*. Washington, D.C.: National Governors Association.

Porter, M. (2000). Location, competition, and economic development: Local clusters in a global economy. *Economic Development Quarterly*, 14(1), 15-34.

Reid, N., Carroll, M., Smith, B. (2007). Critical steps in the cluster building process. *Economic Development Journal*, 6(4), 44–52.

Reid, N., Smith, B., Carroll, M. (2008). Cluster regions: A social network perspective. *Economic Development Quarterly*, 22, 345–352.

Roberts, B., Stimson, R. (1998). Multi-sectoral qualitative analysis: a tool for assessing the competitiveness of regions and formulating strategies for economic development. *Annals of Regional Science*, 32(4), 469–494.

San Francisco Enterprise GIS Program. http://www.sfgov.org/site/gis_index.asp. Accessed 5 September 2008.

Scott, J. (2000). *Social Network Analysis*, 2nd edition, Thousand Oaks, CA: Sage Publications.

SF Prospector. http://www.sfgov.org/site/sfprospector_index.asp. Accessed 5 September 2008.

Smith, B., Carroll, M., Reid, N. (2007). Potential cluster regions; the case of the U.S. floriculture industry. *Papers of the Applied Geography Conferences*, 30, 59–66.

Spencer, J.H., Ong, P. (2004). An analysis of the Los Angeles revitalization zone: Are place-based investment strategies effective under moderate economic conditions? *Economic Development Quarterly*, 18(4), 368–383.

Sui, D. (1995). Spatial economic impacts of new town development in Hong Kong: A GIS-based shift-share analysis. *Socio-Economic Planning Sciences*, 29(3), 227–243.

Texas Industry Profiles (2004). *Texas industry cluster initiative background.* http://www.texasindustryprofiles.com/PDF/twcClusterReports/ClusterInitiativeBackground.pdf. Accessed 5 September 2008.

U.S. Census Bureau (2005) *County business patterns, 2005.* http://www.census.gov/epcd/cbp/view/cbpview.html. Accessed 5 September May 2008.

U.S. Department of Commerce. Bureau of Economic Analysis (2003). *State and local personal income, state annual estimates, 2003.* http://www.bea.gov/regional/rims/. Accessed 5 September 2008.

Wang, F. (2006). *Quantitative methods and applications in GIS.* Boca Raton: Taylor & Francis Group.

Wasserman, S., Faust, K. (1994). *Social network analysis: Methods and applications.* Cambridge: Cambridge University Press.

Watts, D., Wood, A., Wardle, P. (2006). Owner-managers, clusters, and local embeddedness: Small firms in the Sheffield (UK) metal-working cluster. *Entrepeneurship & Regional Development*, 18(3), 185–205.

World Bank (2008). What is LED? www.worldbank.org. Last accessed 5 September 2008.

Chapter 3
Identifying Urban Neighborhoods for Tree Canopy Restoration Through Community Participation

Jeffrey S. Wilson and Greg H. Lindsey

Abstract This chapter describes the development of an urban tree canopy cover assessment and restoration project designed to inform community stakeholders and to guide tree planting efforts in disadvantaged neighborhoods surrounding the Indianapolis, Indiana central business district. The primary goals of the project were to create estimates of tree canopy cover and distribution within the study region, and to identify neighborhoods in which to focus tree planting efforts. Estimates derived from high resolution satellite imagery and aerial photography indicate approximately 17% of the studying region was covered by tree canopy. A site selection model identified eight residential neighborhoods that exhibited social and environmental characteristics prioritized for tree planting. These neighborhoods contained significantly less tree cover than other residential communities. The results of this study were used to inform a tree planting campaign, the Indianapolis NeighborWoods initiative, with a goal of planting 100,000 trees in the study region.

Keywords Urban Forestry · Neighborhood Selection Model · Socioeconomics · Remote Sensing · Indianapolis

Trees provide environmental, economic, and social benefits to cities and their inhabitants (McPherson 1992; Elmendorf 2008). While efforts to manage urban trees in the US date back to the colonial era, comprehensive scientific approaches to urban forestry began to emerge during the 1960s (Konijnendijk 2003; Konijnendijk et al. 2006). Over the last several decades, researchers have sought to increase our understanding of the specific ways trees enhance the quality of urban life and environments. Studies on processes such as carbon sequestration and reduction of air pollution (Nowak 1993; Nowak and Crane 2002; Myeong et al. 2006; Jim and Chen 2008; Nowak et al. 2006), energy conservation and mitigation of the urban heat island effect (McPherson and Simpson 2003; Wilson et al. 2003; Hardin

---

J.S. Wilson (✉)
Department of Geography, Indiana University – Purdue University Indianapolis, Indianapolis, IN, USA
e-mail: jeswilso@iupui.edu

J.D. Gatrell, R.R. Jensen (eds.), *Planning and Socioeconomic Applications*, Geotechnologies and the Environment 1, DOI 10.1007/978-1-4020-9642-6_3,

and Jensen 2007), and the sustainability of urban hydrologic systems (Paul and Meyer 2001; Nowak 2006) continue to advance our knowledge about the importance of urban trees. Related research in health and social sciences suggests urban trees and green space are associated with higher property values (Mansfield et al. 2005; Ellis et al. 2006), increased traffic on pedestrian trails (Lindsey et al. 2006,), more favorable perceptions of neighborhood walking environments (Foltête and Piombini 2007; Liu et al. 2007), improved cognitive functioning and human health (Frumkin 2001; Taylor et al. 2001; Nielsen and Nilsson 2007), and enhanced social interactions (Kuo 2003; Westphal 2003).

Increased recognition and improved understanding of the ways trees enhance life in cities has prompted interventions to increase urban tree canopy cover through planting campaigns. Many urban tree planting projects bring together members from non-profit, government, community, and research sectors. Recent examples of cities planning for or implementing urban forest restoration projects include Addis Ababa (Horst 2006), Beijing (Yang et al. 2005), Los Angeles (McPherson et al. 2008) and Milwaukee (Perkins et al. 2004), but there are many such efforts around the globe that range from national to neighborhood scales. As organizations are faced with decisions about how to manage urban tree canopy restoration, information about existing canopy cover and the spatial distribution of social and environmental characteristics that might benefit from increased tree canopy are needed in order to inform prioritization and distribution of resources.

This chapter describes the development of the first phase of an urban tree canopy assessment and restoration project designed to inform community stakeholders and to guide tree planting efforts in disadvantaged neighborhoods surrounding the Indianapolis, Indiana central business district (CBD). The project involved collaboration between community members organized by Keep Indianapolis Beautiful (KIB) Inc., a local not-for-profit organization, and faculty from the School of Liberal Arts and School of Public and Environmental Affairs at Indiana University – Purdue University Indianapolis (IUPUI). The primary goals of the project were to create current estimates of tree canopy cover and its distribution within the study region, and to identify neighborhoods in which to focus tree planting efforts.

3.1 Background

Remote sensing and GIS technologies are increasingly used to support integration of social and environmental research in cities, including applications in urban forestry (Gatrell and Jensen 2008). High resolution imagery acquired from aerial and satellite platforms are a primary data source for developing maps and estimates of urban tree canopy cover. Tree canopy cover refers to the percentage of ground area covered by foliage in a planimetric view (Nowak et al. 1996). As a metric in urban forest research, canopy cover has been referred to as "the driving force behind the urban forest's ability to produce benefits for the community" (Peper et al. 2008).

Recent studies have explored the use of geospatial technologies specifically for identify tree planting locations for urban canopy restoration and enhancement projects. For example, Wu et al. (2008) integrated impervious surface, land cover, and existing canopy cover data to develop a site selection model for tree planting in Los Angeles, California. Their results identified over 2 million planting sites that have potential to increase total tree canopy cover in the city by about 110 sq km.

In addition to physical environment criteria, site selection for urban tree planting can integrate social values of communities through public participation approaches. Public participation offers opportunities to integrate opinions and perceptions of community members through inclusion in decision making processes. Projects that integrate public participation and GIS methods are commonly referred to in geography literature using the acronym PPGIS (public participation GIS) (Sieber 2006). Studies of public participation in urban forestry suggest that inclusion of community members can improve tree survival rates, enhance relationships between citizens and urban forest experts, and help to leverage funding and political support for urban forest projects (McPherson and Johnson 1988; Austin 2002).

Approaches that have been used to facilitate community involvement in urban forest planning include resident interviews and surveys, public meetings, field trips, and collaborative planning groups (Sipilä and Tyrväinen 2005; Van Herzele et al. 2005). Integrating both physical and social criteria in urban tree canopy restoration facilitates an ecosystem perspective that considers the interaction between a city's inhabitants and the physical environment. Zipperer (2008) suggests this perspective on urban forestry gained popularity in the US during the 1990s in concert with development of broader views of cities as ecosystems and the concept of urban ecosystem services.

The study area for the current project was Center Township in Marion County, Indiana, encompassing an area of approximately 110 sq km (see inset map in Fig. 3.1). The core of the township includes the Indianapolis CBD, the State Capital and Indiana Government Center, university and hospital campuses, and other commercial and cultural venues typical of a large Midwestern capital city. Center Township had a population of 167,055, a median household income of $26,435, and 78,325 total housing units according the 2000 US Census. The township is more densely populated, racially diverse, and has higher proportions of persons in poverty, crime, vacant housing, and renter-occupied housing compared to the city as a whole. A variety of single and multifamily residential communities, some with strong neighborhood identities, surround the urban core of the township, as well as a few industrial complexes. Most of the residential neighborhoods are composed of older housing units (median year structure built = 1944) , but urban renewal and infilling projects have created some new and revitalized housing units in both residential neighborhoods and in mixed-use settings closer to the urban core.

3.2 Methods

3.2.1 Urban Tree Canopy Mapping

Recent tree canopy cover estimates for Center Township were derived using high resolution satellite imagery and digital aerial photography. Satellite imagery collected on April 25, 2005 from the QuickBird satellite was provided through a grant from the Institute for Application of Geospatial Technology. The QuickBird sensor collects multispectral imagery in blue, green, red, and near infrared portions of the electromagnetic spectrum at a spatial resolution of 2.4 m, and panchromatic imagery at a spatial resolution of 60 cm. High resolution aerial photography was used to supplement satellite image interpretation of tree canopy cover. Natural color air photos collected in 2004 at a spatial resolution of 15 cm were provided by the Indianapolis Mapping and Geographic Infrastructure System (IMAGIS) via Indiana University's Spatial Data Portal (http://www.indiana.edu/~gisdata/).

Standard digital image processing techniques were used to group pixels in the multispectral QuickBird satellite imagery into 50 clusters based on similarity in spectral response. Clustering was implemented using the Iterative Self Ordering Data Analysis Technique (ISODATA) unsupervised classification algorithm. The resulting spectral clusters were assigned to one of three categories (tree canopy, other, and mixed) based on visual interpretation of the imagery and the spectral signatures associated with each cluster. The mixed category was composed of spectral clusters containing a mixture of tree canopy and other land cover types. A cluster busting procedure was implemented on pixels in the mixed category by applying additional ISODATA classifications to further separate tree canopy from other land cover types.

Other digital image processing techniques were tested as potential improvements in the classification process, including measures of image texture and merging the multispectral and panchromatic bands of the QuickBird imagery. The results of these enhancements were only marginally successful based on visual comparison of the classified images to high resolution photography. Because the purpose of the study was to create a tree canopy map that could be used in subsequent phases of the project, rather than evaluation of urban tree canopy classification methods, visual interpretation and onscreen digitizing were used to refine the satellite image classification.

The tree canopy map created from ISODATA classification was overlaid on 15 cm aerial photography and a remote sensing analyst visually compared the map with the high resolution photos in a heads-up digitizing environment. Obvious errors resulting from the digital classification process were manually corrected by changing the pixel values to the appropriate category (tree canopy or other) based on interpretation of the high resolution photography. Approximately 75 hours of image analyst time were devoted to manual correction.

A random sample of 100 pixels was used to derive an estimate of the accuracy of the tree canopy map. The accuracy assessment was conducted by comparing the class of the sample points (tree canopy or other) to visual interpretation of the aerial photography. Eighty-nine of the sample points were classified correctly. The relatively high accuracy is likely attributable to the detailed manual corrections that were implemented through visual interpretation. Although accuracy assessment would ideally be conducted independently of the aerial photography used in visual interpretation, this imagery was the best available data for the study region at the time from which to interpret land cover types. Overall, the urban tree canopy map developed for this project was the most up-to-date and accurate source available for the study region.

Table 3.1 Variables used to develop a neighborhood selection model for urban tree canopy restoration in Center Township, Marion County, Indiana

Variable	Description	Source
Median household income	Census block groups with median household income below the township average were coded with a value of 1, otherwise 0.	2000 US Census
Crime	Density of Uniform Crime Report Part 1 and Part 2 crimes in 2004 was calculated by dividing the total number of crimes in census block groups by the block group area. Census block groups with values greater than the township average were coded with a value of 1, otherwise 0.	Social Assets and Vulnerability Indicators (SAVI) Database
Residential zoning	Residential areas delineated by the Indianapolis Department of Metropolitan Development's zoning GIS layer were coded with a value of 1, otherwise 0.	IMAGIS
Industrial emission sites	Locations in Center Township with Title V industrial emission permits were buffered at a distance of 400 meters. Areas within the 400 meter buffers of a Title V emission sites were coded with a value of 1, otherwise 0.	IMAGIS
Pediatric asthma	Pediatric asthma rates by zip code in the year 2000 based on hospital discharge data for children ages 5 to 14 were obtained from the Health and Hospital Corporation of Marion County. Zip code polygons with higher than average rates for the township were given a value of 1, otherwise 0.	Health & Hospital Corporation of Marion County
Major road proximity	Areas within 150 m of major road center lines were coded with a value of 1, otherwise of 0. Major roads include expressways, freeways, primary and secondary arteries.	IMAGIS
Surface temperatures	Radiant surface temperatures in Center Township were estimated from a Landsat 7 Enhanced Thematic Mapper Plus (ETM+) image collected on June 6, 2000. Average radiant temperature within census blocks was calculated using a GIS zonal operation. Census blocks with radiant surface temperatures above the township average were coded with a value of 1, otherwise 0.	Landsat 7 ETM+
Impervious surface cover	Estimates of impervious surface cover were derived from GIS data layers including building footprints, buffers of road center lines, and impervious surface areas digitized by the City of Indianapolis. Percent area covered by impervious surface within census blocks was calculated using a GIS zonal operation. Census blocks with percent impervious surface cover above the township average were coded with a value of 1, otherwise 0.	IMAGIS
Tree canopy cover	Percent area covered by tree canopy within census blocks was calculated using a GIS zonal operation. Areas below the township average were coded with a value of 1, otherwise 0.	QuickBird image and aerial photography

3.2.2 Neighborhood Selection Model

A site selection model is a decision support tool for identifying locations where multiple criteria overlap. The purpose of the site selection model used in this project was to identify specific Center Township neighborhoods in which to focus tree planting efforts that would later be implemented by KIB and their community partners. KIB assembled a collaborative planning group of community leaders to select socioeconomic and environmental characteristics as priorities in the site selection model. This group included representatives from local government, business, utility companies, and neighborhood organizations. Expertise of group members included urban forestry, community organization, business, non-profit leadership, and urban development. Through a series of collaborative planning meetings, group members identified nine social and environmental neighborhood characteristics as priorities for focusing tree planting efforts: residential land use; high rates of crime and pediatric asthma; low income; close proximity to industrial emission sites and major roads; higher radiant surface temperatures and impervious surface cover; and less tree canopy cover relative to the township as a whole.

A primary goal of the model development phase was to use methods that were as straightforward as possible so that the process of selecting neighborhoods could be clearly communicated to the general public. Each of the social and environmental criteria was represented as a binary map layer in a GIS. A value of 1 was used to represent locations exhibiting characteristics prioritized for tree canopy restoration according to the criteria selected by the collaborative planning group, while a value of 0 was applied to other locations. For example, one of the goals of the projects was to increase tree canopy cover in high crime neighborhoods. Areas with crime rates higher than the township average were coded with a value of 1 and areas below the average were given a value of 0. Table 3.1 describes the nine layers incorporated in the neighborhood selection model, their data sources, and method of specification.

All of the model layers were added together to produce the final neighborhood selection map, with values potentially ranging from 0 (no criteria occur at a location) to 9 (all criteria occur at a location). When explaining the processes to lay audiences, KIB and collaborative planning group members referred to the manual transparency overlay methods pioneered by Ian McHarg in *Design with Nature* (1969) to provide an analogy for the neighborhood selection modeling process implemented in a GIS environment.

3.3 Results

3.3.1 Urban Tree Canopy Cover Estimates

The Center Township tree canopy map developed from QuickBird imagery and aerial photography is depicted in Fig. 3.1. Estimates derived from this map indicate approximately 17% of the township (19 of 110 sq km) was covered by tree canopy

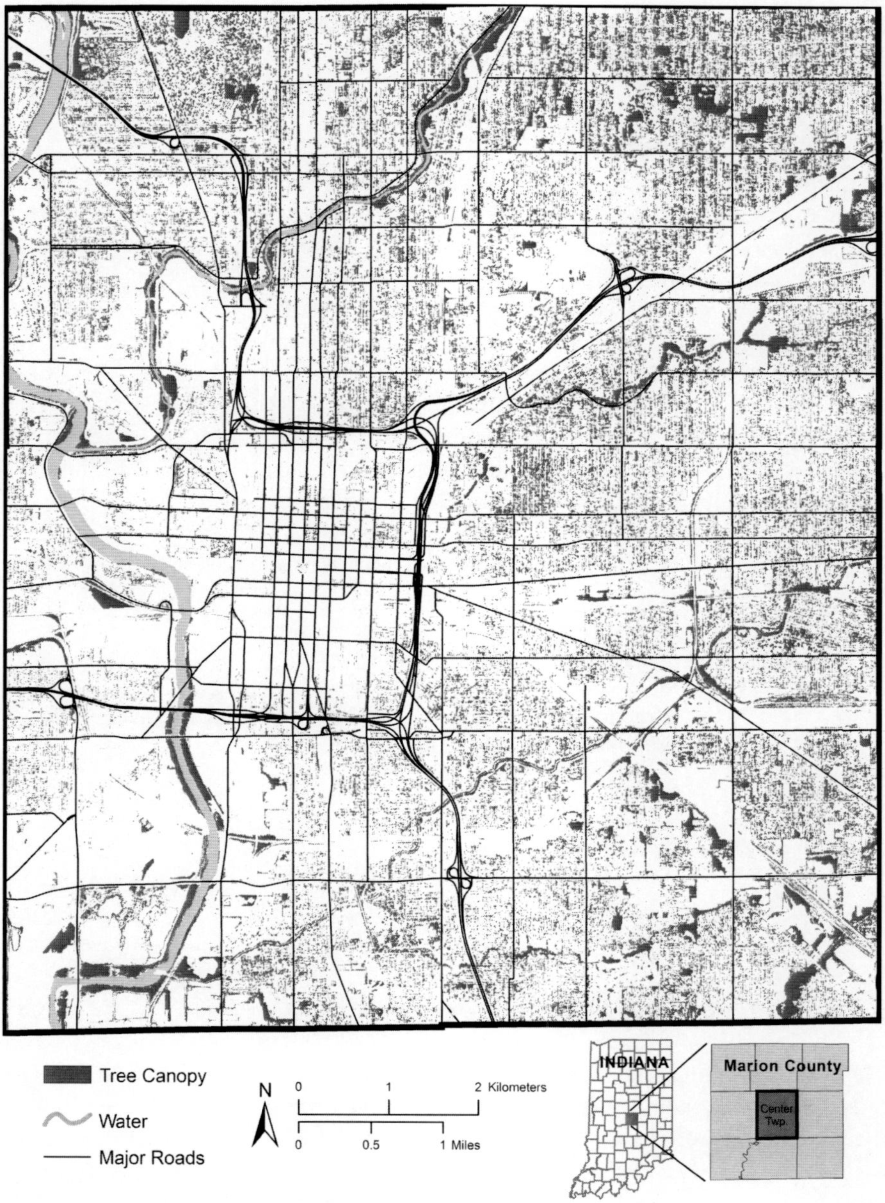

Fig. 3.1 Tree canopy cover in Center Township, Marion County, Indiana

in 2005. Dwyer and Nowak (2000) suggest that tree canopy cover in urban areas of the US averages 27%. While the canopy cover estimates for Center Township were well below this average, the current study focused on the most densely developed portion of Indianapolis. Estimates of canopy cover for the entire city undoubtedly would be higher than those for Center Township alone because there development is less dense and more residential land use and parks occur in areas surrounding the urban core. A recent report by the USDA Forest service estimated that street trees alone create 14% tree canopy cover over the entire city of Indianapolis (Peper et al. 2008).

Land use is primary factor in determining the distribution of tree canopy cover in urban environments (Nowak et al. 1996). Percent tree canopy in Center Township was summarized by general zoning categories in Table 3.2. The highest percent canopy cover occurred in parks, as might be expected. Canopy cover in residential areas of Center Township was second highest at approximately 23%. Center Township's residential areas, however, are slightly below the 25% canopy cover recommended by American Forests for urban residential areas east of the Mississippi River and in the Pacific Northwest (American Forests 2008). The Indianapolis CBD, university, and hospital zoning categories had the lowest percent canopy cover, all at around 4%. American Forest recommends canopy cover of 15% in CBDs, which is more than triple the 2005 estimates for the Indianapolis CBD.

Table 3.2 Percent area covered by tree canopy in general zoning categories, Center Township, Marion County, Indiana

Zoning type	Canopy cover (%)
CBD	4.3
Commercial	8.4
Historic	18.7
Hospital	4.2
Industrial	9.9
Park	25.3
Residential	23.3
Special Use	17.5
University	4.3
Township (total)	17.4

3.3.2 Neighborhood Selection Model

The output of the neighborhood selection model is depicted in Figure 3.2. Darker tones in the map indicate areas where more of the criteria were met, while lighter tones indicate fewer criteria occurred at a location. Eight hot spots (areas where most or all of the model criteria occurred) emerged from the modeling process. Table 3.3 provides percent canopy cover estimates for the eight hot spot neighborhoods. Hot spot neighborhoods had on average 9% less tree canopy cover when compared to all residential areas in the township. While residential areas in the township as a whole approached the 25% tree canopy cover recommended by American Forests, Inc., all of the hot spot neighborhoods were well below this recommended level.

Table 3.4 presents calculations estimating the number of additional trees that would be necessary to achieve 25% canopy cover for urban residential areas in the eight Center Township hot spot neighborhoods and 15% canopy coverage in the

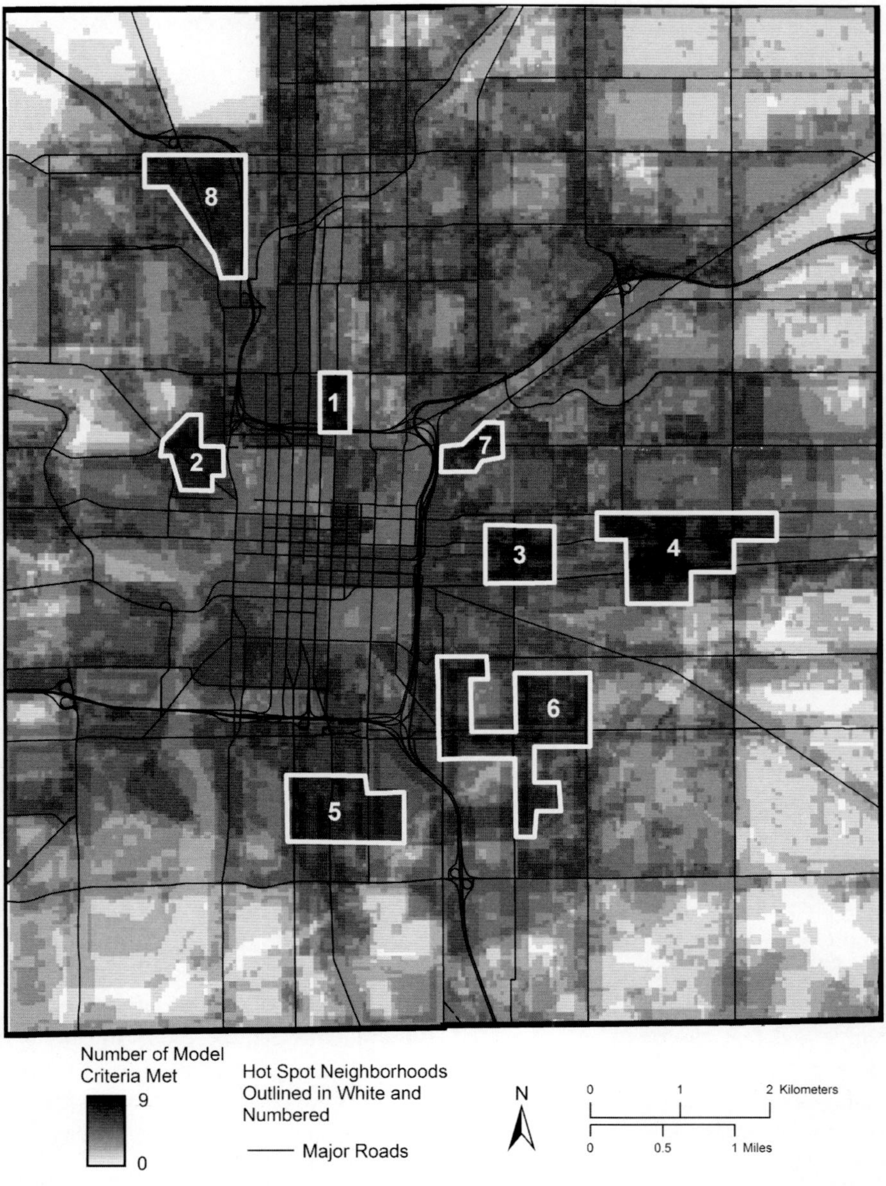

Fig. 3.2 Output of neighborhood selection model for urban tree canopy restoration showing hot spots in Center Township, Marion County, Indiana

CBD. These calculations make are based on some broad assumptions, but serve as a starting point for more detailed planting strategies. In these calculations it is assumed that current canopy is held constant, that all trees planted survive to develop a canopy coverage area of about 12 sq m (the average area of urban trees

Table 3.3 Percent area covered by tree canopy in hot spot neighborhoods, Center Township, Marion County, Indiana

Region	Canopy cover (%)
Hot Spot 1	13.8
Hot Spot 2	9.6
Hot Spot 3	15.4
Hot Spot 4	15.5
Hot Spot 5	11.0
Hot Spot 6	16.6
Hot Spot 7	17.0
Hot Spot 8	15.3

Table 3.4 Estimates trees needed to achieve recommended canopy cover levels in hot spot neighborhoods and the Indianapolis CBD, Center Township, Marion County, Indiana

Region	Total area (m^2)	Current canopy cover (m^2)	Current canopy cover (%)	Target canopy cover (%)	Avg. cover per tree (m^2)	Trees to achieve target canopy cover
Hot spot 1	189,344	26,037	13.75	25	12.36	1,724
Hot spot 2	303,485	29,265	9.64	25	12.36	3,772
Hot spot 3	479,766	73,835	15.39	25	12.36	3,731
Hot spot 4	1,247,095	193,113	15.49	25	12.36	9,603
Hot spot 5	881,345	97,380	11.05	25	12.36	9,951
Hot spot 6	1,227,450	203,655	16.59	25	12.36	8,353
Hot spot 7	186,454	31,673	16.99	25	12.36	1,209
Hot spot 8	770,426	116,853	15.17	25	12.36	6,131
					Hot spot subtotal	44,474
CBD	4,541,983	195,305	4.30	15	12.36	39,332
					Total	83,806

reported by American Forests, Inc.), and that individual tree canopies do not overlap. Under these assumptions, approximately 45,000 more trees would be needed in order to meet the 25% canopy cover recommendation in hot spot neighborhoods. Similar calculations are presented at the bottom of Table 3.4 to estimate the number of trees needed to meet the recommended 15% canopy cover in the Indianapolis CBD. Collectively, these calculations indicate that approximately 84,000 trees in the hot spot neighborhoods and CBD would be needed to raise tree canopy cover to recommended levels.

3.4 Summary and Conclusions

An urban tree canopy cover map was developed for Center Township in Marion County, Indiana, which encompasses the urban core of the city of Indianapolis. Canopy cover estimates derived from high resolution satellite imagery and aerial photography indicate that both the CBD and surrounding urban residential areas fall below recommend canopy cover levels. A site selection model identified eight neighborhoods within the study region that exhibited social and environmental characteristics prioritized for tree planting by a collaborative group of community stakeholders. These neighborhoods contained significantly less tree cover than other residential communities in the study region. Rough estimates indicate approximately 84,000 more trees would be needed to meet recommended urban canopy cover levels within residential and CBD areas of the study region.

The neighborhood selection variables implemented in this study were derived by consensus from a collaborative planning group through what was essentially a political process. Some members of the planning group possessed expertise in urban forestry, but the majority were not research scientists or urban forest experts. Thus, while informed by the scientific literature on urban forestry, the criteria used in the model reflect the values and opinions of the planning group participants. The collaborative planning group approach to public participation implemented in this study was successful in collecting input from a group made up of individuals who are probably best described as active community members or leaders. Tyrväinen et al. (2007) point out the challenge of integrating the perspective of 'silent groups' (those who choose not to participate in public forums) in urban forestry projects. Future applications of the methods used in this study could potentially generate more input from neighborhood residents using other methods to facilitate community participation, including resident surveys and interviews.

The methods used in this project have several limitations. The neighborhood selection model was intentionally simplistic. One of the goals of the project was to create a method for selecting neighborhoods that could be communicated easily to non-experts. This goal influenced decisions on the model's complexity and more sophisticated techniques for neighborhood selection could result in different outcomes. For example, all of the social and environmental criteria were given equal weight in the model. Future studies could explore weighting of variables based on expert opinions or community priorities. Each variable was reduced to a binary form based on average conditions in the study region, but more complex approaches could explore ordinal or continuous variable specification, or the use of different thresholds. Additionally, the general criteria integrated in the model could change as a result of including different members in the collaborative planning group. The neighborhoods identified in the modeling process would likely benefit the from community tree planting efforts, but finer scale analyses could be conducted to identify specific locations where individual trees can be planted within these neighborhoods. The urban forest literature provides recommendations about specific site characteristics and tree species desirable for tree planting projects, e.g. Cappiella et al. (2006).

Subsequent to the planning work described in this chapter, KIB and their community partners have taken the important step of beginning the processes of planting trees. The planting campaign, referred to as the Indianapolis NeighborWoods initiative, has a stated goal of planting 100,000 trees in the Indianapolis CBD and hot spot neighborhoods identified in this study over a ten year period from 2007 to 2017. Community reactions to the Indianapolis NeighborWoods initiative have been highly supportive, including positive local media coverage, financial and political support, and community-organized tree planting events. As of the summer of 2008, approximately 6,000 trees have been planted in the hot spot neighborhoods identified through the research described in this chapter. It is hoped that the trees planted and those to come will contribute to enhancing the quality of urban life in Indianapolis.

References

American Forests, Inc. (2008). Setting *Urban Tree Canopy Goals*. http://www.americanforests.org/resources/urbanforests/treedeficit.php. Accessed 15 August 2008.

Austin, M. E. (2002). Partnership opportunities in neighborhood tree planting initiatives: building from local knowledge. *Journal of Arboriculture*, 28(4), 178–186.

Cappiella, K., Schueler, T., Tomlinson, J., & Wright, T. (2006). *Urban Watershed Forestry Manual Part 3: Urban Tree Planting*. http://www.na.fs.fed.us/watershed/pdf/Urban Watershed Forestry Manual Part 3.pdf. Accessed 15 August 2008.

Dwyer, J. F., & Nowak, D. J. (2000). A national assessment of the urban forest: An overview. Paper presented at the Society of American Foresters 1999 National Convention, Portland, Oregon.

Ellis, C. D., Lee, S. W., & Kweon, B. S. (2006). Retail land use, neighborhood satisfaction and the urban forest: An investigation into the moderating and mediating effects of trees and shrubs. *Landscape and Urban Planning*, 74(1), 70–78.

Elmendorf, W. (2008). The importance of trees and nature in community: A review of the relative literature. *Arboriculture & Urban Forestry*, 34(3), 152–156.

Foltête, J.-C., & Piombini, A. (2007). Urban layout, landscape features and pedestrian usage. Landscape and Urban Planning, 81(3), 225–234.

Frumkin, H. (2001). Beyond toxicity: Human health and the natural environment. *American Journal of Preventive Medicine*, 20(3), 234–240.

Gatrell, J. D., & Jensen, R. R. (2008). Sociospatial applications of remote sensing in urban environments. *Geography Compass*, 2(3), 728–743.

Hardin, P. J., & Jensen, R. (2007). The effect of urban leaf area on summertime urban surface kinetic temperatures: A Terre Haute case study. *Urban Forestry & Urban Greening*, 6(2), 63–72.

Horst, A. (2006). Rehabilitation of urban forests in Addis Ababa. *Journal of Drylands*, 1(2), 108–117.

Jim, C. Y., & Chen, W. Y. (2008). Assessing the ecosystem service of air pollutant removal by urban trees in Guangzhou (China). *Journal of Environmental Management*, 88(4), 665–676.

Konijnendijk, C. C. (2003). A decade of urban forestry in Europe. *Forest Policy and Economics*, 5(2), 173–186.

Konijnendijk, C. C., Ricard, R. M., Kenney, A., & Randrup, T. B. (2006). Defining urban forestry - A comparative perspective of North America and Europe. *Urban Forestry & Urban Greening*, 4(3-4), 93–103.

Kuo, F. E. (2003). The role of arboriculture in healthy social ecology. *Journal of Arboriculture*, 29(3), 148–155.

Lindsey, G. H., Han, Y. L., Wison, J. S., & Yang, J. H. (2006). Neighborhood correlates of urban trail use. *Journal of Physical Activity and Health*, 3(S1), S139–S157.

Lindsey, G. H., Wilson, J. S., Yang, J. H., & Alexa, C. P. (2008). Urban form, trail characteristics, and trail use: Implications for urban design. *Journal of Urban Design*, 13(1), 53–79.

Liu, G. C., Colbert, J. T., Wilson, J. S., Yamada, I., & Hoch, S. C. (2007). Examining urban environment correlates of childhood physical activity and walkability perception with GIS and remote sensing. In R. R. Jensen, J. D. Gatrell & D. D. McLean (Eds.), *Geo-Spatial Technologies in Urban Environments: Policy, Practice, and Pixels, 2nd Edition* (pp. 242). Berlin: Springer-Verlag.

Mansfield, C., Pattanayak, S. K., McDow, W., McDonald, R., & Halpin, P. (2005). Shades of green: Measuring the value of urban forests in the housing market. *Journal of Forest Economics*, 11(3), 177–199.

McHarg, I. L. (1969). *Design with Nature*. Garden City, New York: Doubleday / Natural History Press.

McPherson, E. G. (1992). Accounting for benefits and costs of urban greenspace. *Landscape and Urban Planning*, 22(1), 41–51.

McPherson, E. G., & Johnson, C. W. (1988). A community forestry planning process: Case study of citizen participation. *Landscape and Urban Planning*, 15(1–2), 185–194.

McPherson, E. G., & Simpson, J. R. (2003). Potential energy savings in buildings by an urban tree planting programme in California. *Urban Forestry & Urban Greening*, 2(2), 73–86.

McPherson, E. G., Simpson, J. R., Xiao, Q., & Chunxia, W. (2008). *Los Angeles 1-Million tree canopy cover assessment* (No. PSW-GTR-207). Albany, CA: USDA Forest Service, Pacific Southwest Research Station.

Myeong, S., Nowak, D. J., & Duggin, M. J. (2006). A temporal analysis of urban forest carbon storage using remote sensing. *Remote Sensing of Environment*, 101(2), 277–282.

Nielsen, A. B., & Nilsson, K. (2007). Urban forestry for human health and wellbeing. *Urban Forestry & Urban Greening*, 6(4), 195–197.

Nowak, D. J. (1993). Atmospheric carbon reduction by urban trees. *Journal of Environmental Management*, 37(3), 207–217.

Nowak, D. J. (2006). Institutionalizing urban forestry as a "biotechnology" to improve environmental quality. *Urban Forestry & Urban Greening*, 5(2), 93–100.

Nowak, D. J., & Crane, D. E. (2002). Carbon storage and sequestration by urban trees in the USA.. *Environmental Pollution*, 116(3), 381–389.

Nowak, D. J., Crane, D. E., & Stevens, J. C. (2006). Air pollution removal by urban trees and shrubs in the United States. *Urban Forestry & Urban Greening*, 4(3–4), 115–123.

Nowak, D. J., Rowntree, R. A., McPherson, E. G., Sisinni, S. M., Kerkmann, E. R., & Stevens, J. C. (1996). Measuring and analyzing urban tree cover. *Landscape and Urban Planning*, 36(1), 49–57.

Paul, M. J., & Meyer, J. L. (2001). Streams in the urban landscape. *Annual Review of Ecology and Systematics*, 32(1), 333–365.

Peper, P. J., McPherson, E. G., Simpson, J. R., Vargas, K. E., & Xiao, Q. (2008). City of Indianapolis, Indiana Municipal Forest Resource Analysis: Center for Urban Forest Research, USDA Forest Service, Pacific Southwest Research Station. Document Number)

Perkins, H. A., Heynen, N., & Wilson, J. (2004). Inequitable access to urban reforestation: the impact of urban political economy on housing tenure and urban forests. *Cities*, 21(4), 291–299.

Sieber, R. (2006). Public Participation Geographic Information Systems: A Literature Review and Framework. *Annals of the Association of American Geographers*, 96(3), 491–507.

Sipilä, M., & Tyrväinen, L. (2005). Evaluation of collaborative urban forest planning in Helsinki, Finland. *Urban Forestry & Urban Greening*, 4(1), 1–12.

Taylor, A. F., Kuo, F. E., & Sullivan, W. C. (2001). Coping with ADD: The Surprising Connection to Green Play Settings. *Environment and Behavior*, 33(1), 54–77.

Tyrväinen, L., Mäkinen, K., & Schipperijn, J. (2007). Tools for mapping social values of urban woodlands and other green areas. *Landscape and Urban Planning*, 79(1), 5–19.

Van Herzele, A., De Clercq, E. M., & Wiedemann, T. (2005). Strategic planning for new woodlands in the urban periphery: through the lens of social inclusiveness. *Urban Forestry & Urban Greening*, 3(3–4), 177–188.

Westphal, L. M. (2003). Urban greening and social benefits: a study of empowerment outcomes. *Journal of Arboriculture*, 29(3), 137–148.

Wilson, J. S., Clay, M., Martin, E., Stuckey, D., & Vedder-Risch, K. (2003). Evaluating environmental influences of zoning in urban ecosystems with remote sensing. *Remote Sensing of Environment*, 86(3), 303–321.

Wu, C., Xiao, Q., & McPherson, E. G. (2008). A method for locating potential tree-planting sites in urban areas: A case study of Los Angeles, USA. *Urban Forestry & Urban Greening*, 7(2), 65–76.

Yang, J., McBride, J., Zhou, J., & Sun, Z. (2005). The urban forest in Beijing and its role in air pollution reduction. *Urban Forestry & Urban Greening*, 3(2), 65–78.

Zipperer, W. C. (2008). Applying ecosystem management to urban forestry. In M. M. Carreiro, Y.-C. Song & J. Wu (Eds.), *Ecology, Planning, and Management of Urban Forests: International Perspectives* (pp. 97–108). New York: Springer.

Chapter 4
The Spatially Varying Relationship Between Local Land-Use Policies and Urban Growth: A Geographically Weighted Regression Analysis

Robert Hanham, Richard J. Hoch and J. Scott Spiker

Abstract This chapter examines the geography of urban growth and public policy. The chapter uses geographically weighted regression to investigate observed urbanization and sprawl across southwest Pennsylvania. Using Landsat data, the chapter focuses on observed land use change and the relationship, if any, between change and land use policy. This research shows that the spatial fragmentation of local land-use policy has a variable impact on observed land-use from non-forested open space to an urban built environment.

Keywords Geographically weighted regression (GWR) · Urban sprawl · Remote sensing · Land use policy · Pennsylvania

4.1 Introduction

Accelerating urban growth has generally been viewed as an environmental and social problem in the United States for many years. Recent concerns over global warming and the rising cost of petroleum will only heighten this concern. Sprawl refers to urban growth beyond the outer fringe of an urban area in the surrounding rural area (Clawson 1962; Mills 1981; Peiser 1989). It involves the change from a non-urban land-use to either a low density urban land-use such as housing or a higher density urban land-use such as shopping malls and factories. Urban sprawl tends to be discontinuous, suburban-style development resulting from rapid, apparently unplanned or uncoordinated growth. Eighty percent of the metropolitan areas in the US exhibited increasing sprawl during the 1990s (Lopez and Hynes 2003). Approximately one in five counties in the US is projected to have either increasing or sustained sprawl by 2025 (Burchell et al. 2002).

Political or governmental fragmentation in metropolitan areas has been identified in the literature as one reason for sprawling urban growth (Razin and Rosentraub 2000). Carruthers and Ulfarsson (2002) found that fragmentation is associated with

R. Hanham (✉)
Department of Geology and Geography, West Virginia University, Morgantown, WV, USA
e-mail: robert.hanham@mail.wvu.edu

J.D. Gatrell, R.R. Jensen (eds.), *Planning and Socioeconomic Applications*, Geotechnologies and the Environment 1, DOI 10.1007/978-1-4020-9642-6_4,

lower densities and higher property values. Carruthers (2003) showed that municipal fragmentation is one of several related factors that can lead to disproportionate growth at the urban fringe. Brueckner and Kim (2003) and Daniels (2001) examined the impact that variations of property tax rates among local municipalities within a metropolitan area have on urban growth.

This chapter argues that it is not the fragmentation of local government per se that causes sprawl, but rather the fragmentation of the land-use planning process in an urban area and its hinterland that is responsible for it. Urban land-use planning policy is the most direct regulatory measure used in the facilitation or interference of urban land-use. Metropolitan areas that are politically fragmented and rely on local land-use policy to enact planning initiatives disrupt and fragment the planning process. Fragmented municipal land-use policies are more likely to lead to variable land-use change and uneven urban growth patterns. Various attempts have been made to evaluate the effect of overarching public policies, such as Smart Growth initiatives (Carruthers 2002) or urban containment policies (Wassmer 2006) on urban growth in the US, but no studies could be readily found that have examined the linkage between the fragmentation of land-use planning policy and urban growth.

This aim of this chapter is to identify the relationship between local land-use planning policy and urban growth in and around the Pittsburgh Metropolitan Statistical Area in southwest Pennsylvania, a region characterized by a great deal of sprawl. Local government is highly fragmented in this region, there being more than five hundred local municipalities. This has resulted in a highly fragmented land-use planning process. We use geographically weighted regression (GWR) to estimate the relationship between a number of local land-use planning tools and urban growth throughout the region. The advantage of using GWR in this context is to identify whether this relationship varies spatially. In other words, GWR is designed to examine the spatial structure of non-stationary spatial processes. Our results show that the relationship between local land-use planning policy and urban growth varies considerably across space in the study region, thus supporting our argument that the fragmentation of the local land-use planning process is to some degree responsible for sprawling urban growth.

In the second section of the chapter, we describe the study region. The third section is devoted to a discussion of GWR. The fourth section describes the data and methods used in the research. The fifth section discusses the results of the GWR analysis. The concluding section presents a brief critique of the research and offers some thoughts on the future direction it might take.

4.2 The Study Region

The study region for this research is shown in Fig. 4.1. The region comprises nine counties and 525 local municipalities in southwestern Pennsylvania. The map also shows the location of the Pittsburgh Metropolitan Statistical Area. The population of the study area in 2000 was a little over two and a half million people.

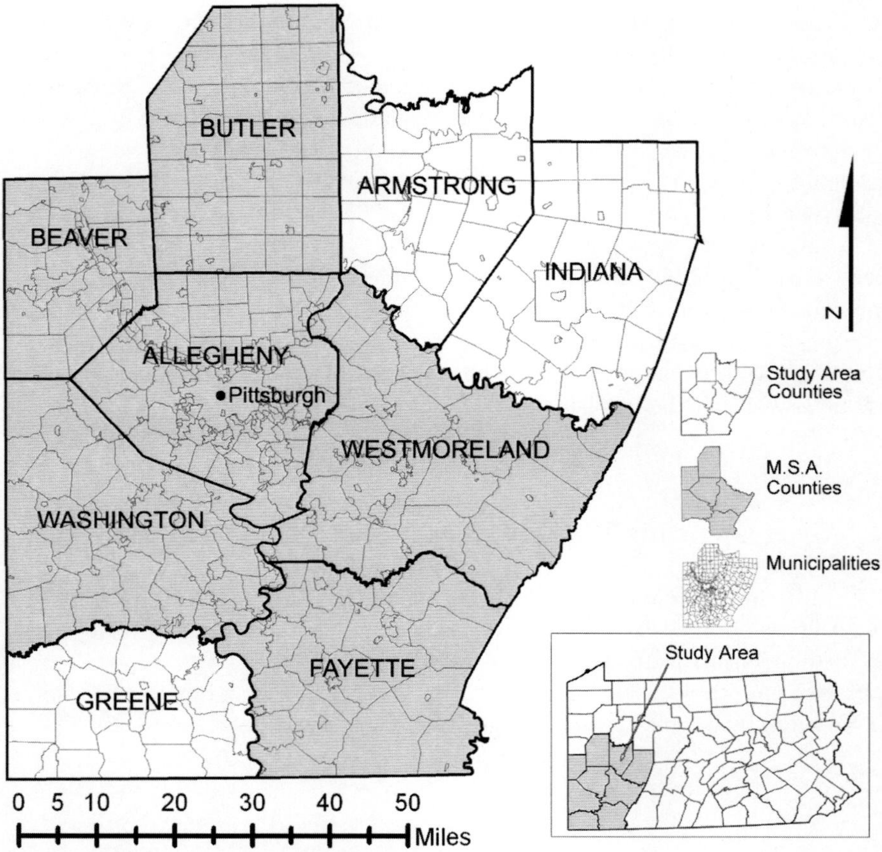

Fig. 4.1 Study region – the nine counties of southwestern Pennsylvania

According to a recent Brookings Institution Center on Urban and Metropolitan Policy report, "Among the nation's largest metropolitan regions, no area can compete with Pittsburgh for profligate land consumption per household" (2003, 47). The report goes on to point out that metropolitan Pittsburgh grew by 202,000 acres from 1982 to 1997, but gained just 24,000 households. This means that the region urbanized 8.4 acres for every new household it added over these 15 years, compared to a national average of 1.3 acres per household. The Brookings report concludes that, "This makes Pittsburgh by far the worst-sprawling large metropolitan area in the country" (2003, 49).

Pittsburgh also has one of the most fragmented local government structures in the US. There are 18 local municipal governments (cities, townships and boroughs) per 100,000 people in the Pittsburgh Metropolitan Statistical Area (MSA). The Pittsburgh MSA has more municipal governments per capita than all other metropolitan areas in the country with populations of at least one million people, of which there are sixty (Briem 1997).

The Southwestern Pennsylvania Commission (SPC), the metropolitan planning organization for the Pittsburgh area, provides planning guidance to the nine counties of the region. However, there are 525 local municipalities in the region, each one tasked with local community, conservation and economic development planning. Each municipality can decide to enact or not to enact local land-use planning policy. Although local municipal governments are empowered to adopt land-use planning tools by the Pennsylvania Municipal Planning Code (MPC), they are not mandated to do so by the state government. This mosaic of local governments operates independently, with little or no cooperation or communication with their neighboring municipalities. Two recent studies of the utilization of land-use planning policies in Pennsylvania, which surveyed local government officials and land-use planning professionals throughout the state, both concluded that local government fragmentation is a barrier to better planning decisions (Lembeck and Kelsey 2001; Lembeck et al. 2001).

4.3 Geographically Weighted Regression

GWR is founded on three interrelated principles (Brunsdon et al. 1996). First, spatial data are not necessarily stationary. Second, the spatial structure of data affects the estimation of relationships based on these data. Third, relationships between variables are not necessarily global, but may be localized and vary across space. GWR reformulates the traditional regression model given by

$$Y_i = \beta_0 + \beta_1 X_i + e_i \tag{4.1}$$

where Y_i and X_i are the dependent and independent variables at point i respectively, β_0 and β_1 are parameters to be estimated and e_i is an error term at point i. The reformulated GWR model is

$$Y_i = \beta_0(u_i, v_i) + \beta_1(u_i, v_i)X_i + e_i \tag{4.2}$$

where (u_i, v_i) represents the coordinates of the i'th data point, and both β_0 and β_1 are continuous functions of (u, v) at point i (Fotheringham et al. 2000).

Operationally, GWR software estimates a separate local equation centered on each data point (Charlton et al. 2002). In each case, the observations are spatially weighted according to their proximity to the central data point, so that observations closer to the central point are given more weight in the estimation than ones further away. The weights are typically expressed as an inverse function of distance from the relevant central point, although the weighting procedure can be specified in various other ways. A spatial kernel is placed over each data point in order to limit the number of observations used to estimate each local equation. GWR allows the user to specify either a fixed or adaptive spatial kernel. Fixed kernels are most suited to situations where data are regularly spaced, and are determined by a fixed spatial limit to the observations that can be used in the estimation of the model for each data point. Adaptive kernels are more suited to situations in which the data are unevenly distributed across space, and are differentially specified for each data point such that the number of observations used in each case is comparable.

GWR provides estimates of the parameters of each local equation, together with local versions of goodness-of-fit measures such as r^2 and the residuals (e_i). Estimates, diagnostics, and residuals can all be mapped. An F test can be used to determine whether the GWR estimates are a significant improvement on the traditional globally estimated equation. Finally, when the areal size of the observations varies, each observation in the regression model can weighted in terms of its area (km^2), an option available in the GWR software.

4.4 Data and Methods

Results from an image-differencing change detection analysis were used to identify the change from non-forested open-space (NFOS) to low-density built environment (LDBE) and high-density built environment (HDBE) in the study area between 1992 and 2002. The change detection involved supervised classifications of Landsat TM and Landsat ETM images (row 17, path 32) dated October 2, 1992 and October 6, 2002 respectively. The supervised classification of both data sets utilized a parametric, maximum likelihood set of decision rules with the use the probabilities assigned to the signatures. The overall classification accuracy for the 1992 and 2002 supervised classifications was 82% in both cases. The Kappa coefficient was 0.78 for the 1992 image and 0.77 for the 2002 image (Hoch, 2005).

The analysis reveals that 17.4% of NFOS in 1992 was transformed into LDBE by 2002, and that 1.1% of NFOS became HDBE by 2002. The resulting cover changes were then assigned to the 525 municipalities in the study area. All image analysis was performed using the Erdas Imagine software package.

The relationship between urban growth and local land-use planning policy in southwest Pennsylvania was examined using a GWR. The GWR model enables us to determine the spatially varying relationship between urban growth and local planning policy. The model is as follows:

$$U_i = \beta_0(u_i, v_i) + \beta_1(u_i, v_i)C_i + \beta_2(u_i, v_i)Z_i + \beta_3(u_i, v_i)S_i + \beta_4(u_i, v_i)P_i + e_i \quad (4.3)$$

where U_i is the percent change in land cover from non-forested open space to urban land-use in municipality i from 1992 to 2002, and C_i, Z_i, S_i and P_i are binary dummy variables which indicate whether or not a municipality has a comprehensive plan, a zoning ordinance, a subdivision regulation and a planning commission respectively.

The GWR model was estimated for two different dependent variables, one representing the change in land cover from non-forested open space to low density urban built environment and the other representing the change in land cover from non-forested open space to high density urban built environment. The data on municipal planning tools was obtained from the Center for Rural Pennsylvania (2001).

Fifty-six percent of municipalities in the region have a comprehensive plan, 61% have a zoning ordinance, 53% have a subdivision regulation and 63% have a planning commission. Fifteen municipalities had to be excluded from the estimation of the two GWR models because of incomplete data. The GWR analysis was therefore based on data for 510 municipalities.

The estimated values of β_1 through β_4 in equation (4.3) will be positive if the presence of a planning tool is associated with more, rather than less, change to urban land cover, and conversely if the planning tool is absent. The parameter estimates will be negative if the presence of a planning tool is associated with less, rather than more, change to urban land cover, and conversely if the planning tool is absent. In other words, the presence of a planning tool is associated with more local urban growth in the former case and is associated with less local urban growth in the latter case. Both GWR models were estimated using an adaptive kernel because of the uneven spatial distribution of municipalities. Lastly, given the wide range of sizes of the municipalities, the observation-weighted version of GWR was used, where each municipality observation was weighted in terms of its area.

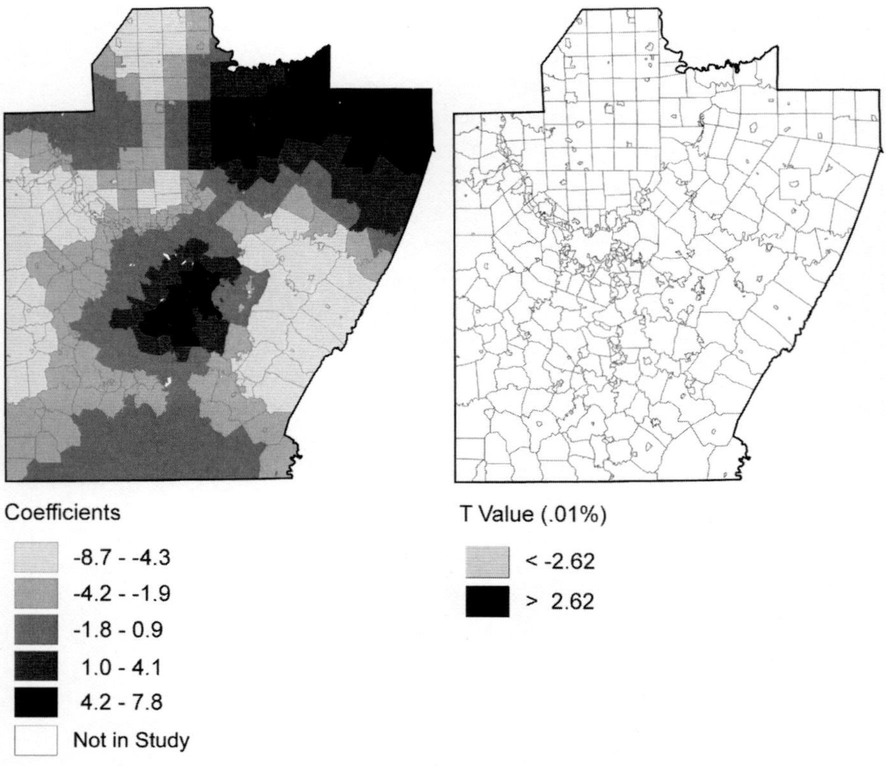

Fig. 4.2 GWR estimate of effect of a Comprehensive Plan on change from non-forested open-space to low-density urban-built environment

4.5 Results

The adjusted r-square for the GWR model of change in land cover from non-forested open space to low density urban built environment is 26 percent. An analysis of variance test ($F = 12.2$) showed that the GWR model is a significant improvement over the global version of the model at the 0.01 level. The local estimates of $\beta_1(u_i, v_i)$ for comprehensive plan, $\beta_2(u_i, v_i)$ for zoning ordinance, $\beta_3(u_i, v_i)$ for subdivision regulation and $\beta_4(u_i, v_i)$ for planning commission are mapped in Figs. 4.2, 4.3, 4.4 and 4.5. Next to each of these maps is a corresponding map of the t-values of those coefficients which are significant at the 0.01 level.

Figures 4.2, 4.3, 4.4 and 4.5 show that the coefficients for all four planning tools vary considerably across the region, suggesting that the effect of the tools on the change in land cover from non-forested open space to low density urban built environment varies considerably across space. However, the corresponding t-values indicate that the variation in the coefficients is much more limited than the impression one gets from the coefficient maps. Figure 4.2 shows that a comprehensive plan is not significant in any part of the region. Figure 4.3 shows that a zoning

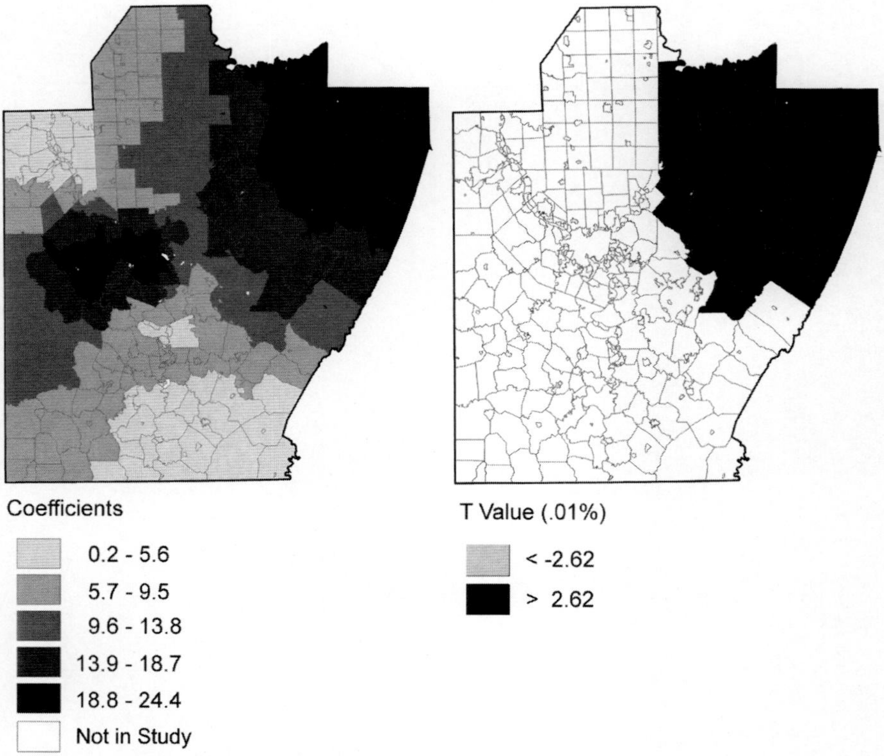

Fig. 4.3 GWR estimate of effect of a Zoning Ordinance on change from non-forested open-space to low-density urban-built environment

ordinance is significant in the northeast quadrant of the region in Armstrong and Indiana counties, and it is insignificant everywhere else. The coefficient is positive throughout the northeast (see Fig. 4.3), indicating that the presence of a zoning ordinance tends to increase the change from non-forested open space to low density urban built environment more than its absence does. Furthermore, the effect of a zoning ordinance on this land cover change intensifies as one moves toward the northeast.

Figure 4.4 shows that a subdivision regulation is significant in a corridor of municipalities in the southeast part of the region, primarily in Fayette County, and in two isolated municipalities to the north in Butler County, and is insignificant everywhere else. The coefficient is negative in all municipalities where it is significant, indicating that the presence of a subdivision regulation tends to decrease the change from non-forested open space to low density urban built environment more than its absence does. Where the effect of a subdivision regulation is significant, however, its effect on this land cover change is uniform across space (compare the coefficient and t-value maps in Fig. 4.4).

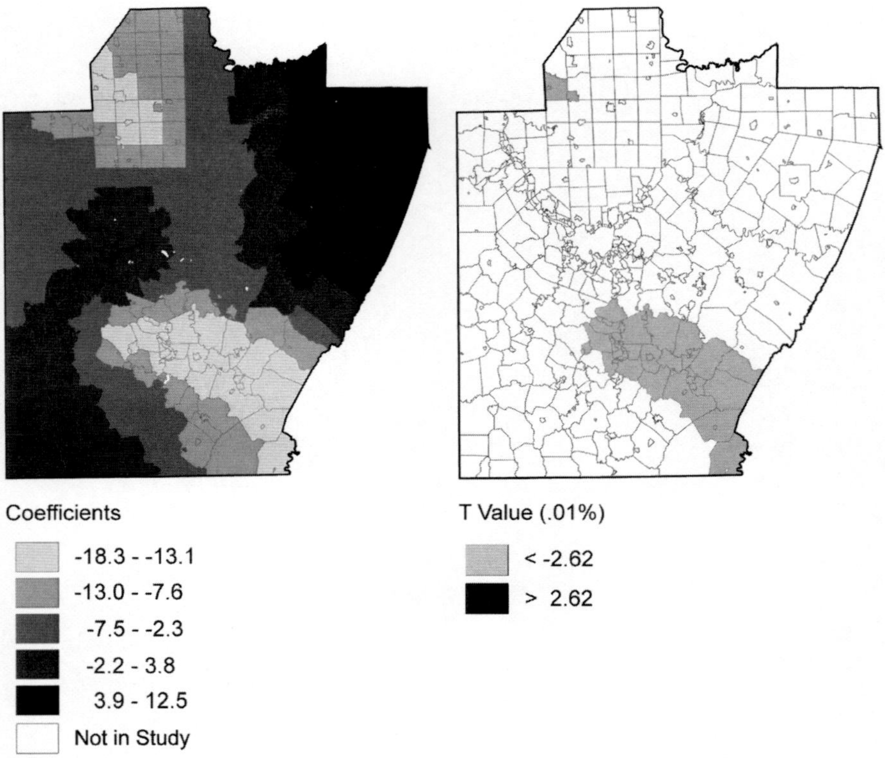

Fig. 4.4 GWR estimate of effect of a Subdivision Regulation on change from non-forested open-space to low-density urban-built environment

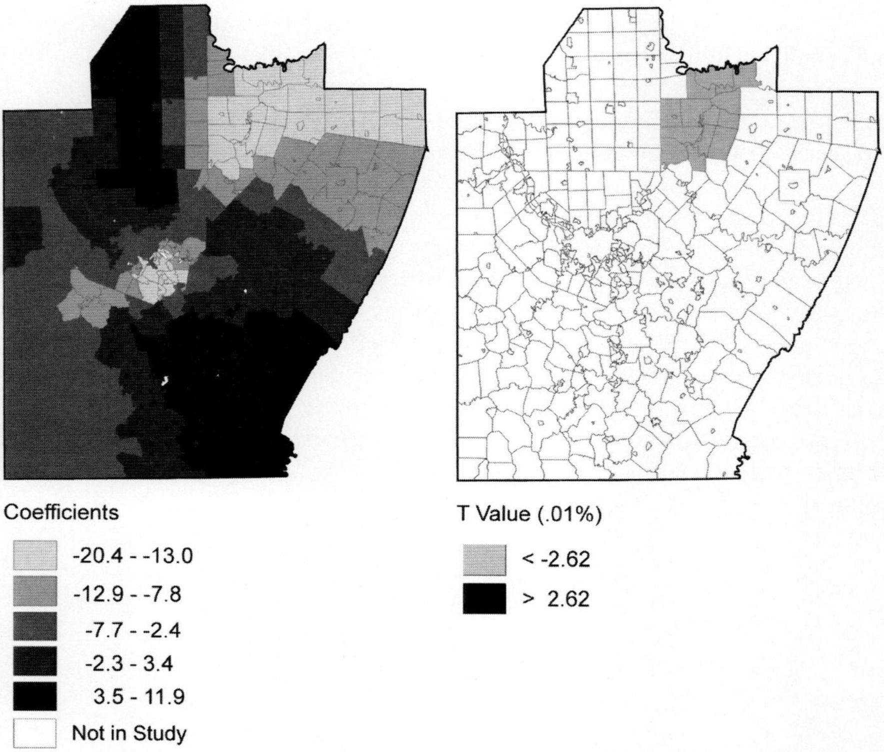

Fig. 4.5 GWR estimate of effect of a Planning Commission on change from non-forested open-space to low-density urban-built environment

Figure 4.5 shows that a planning commission is significant in a small cluster of municipalities in the northern part of the region in Armstrong County, and is insignificant everywhere else. The coefficient is negative throughout this cluster of significant municipalities, indicating that the presence of a planning commission tends to decrease the change from non-forested open space to low density urban built environment more than its absence does. As in the subdivision case, where the effect of a planning commission is significant, its effect on this land cover change is uniform across space (compare the coefficient and t-value maps in Fig. 4.5).

The adjusted r-square for the GWR model of change in land cover from non-forested open space to high density urban built environment is 6 percent. An analysis of variance test ($F = 3.4$) showed that the GWR model is a significant improvement over the global version of the model at the 0.01 level. Despite the very low r-square, the GWR model is significant because of the large number of degrees of freedom. The local estimates of $\beta_1(u_i, v_i)$ for comprehensive plan, $\beta_2(u_i, v_i)$ for zoning ordinance, $\beta_3(u_i, v_i)$ for subdivision regulation and $\beta_4(u_i, v_i)$ for planning commission are mapped in Figs. 4.6, 4.7, 4.8 and 4.9. Next to each of these maps is a corresponding map of the t values of those coefficients which are significant at the 0.01 level.

As was the case for the change from non-forested open space to low density urban built environment, Figs. 4.6, 4.7, 4.8 and 4.9 show that the coefficients for all four planning tools vary considerably across the region with respect to the change in land cover from non-forested open space to high density urban built environment. However, the corresponding t-values shown in Figs. 4.6, 4.7, 4.8 and 4.9 indicate that only the coefficient for subdivision regulation varies significantly across space (see Fig. 4.8). Comprehensive plan, zoning regulation and planning commission are all insignificant throughout the region (see Figs. 4.6, 4.7 and 4.9). Figure 4.8, on the other hand, shows that a subdivision regulation is significant in most of Allegheny county (excluding the eastern part of the county) and in northern Washington county, but is insignificant everywhere else. The coefficient is negative in this area (see Fig. 4.8), indicating that the presence of a subdivision regulation tends to decrease the change from non-forested open space to high density urban built environment more than its absence does. Furthermore, the effect of a subdivision regulation on this land cover change diminishes as one moves outward from the center of Allegheny County.

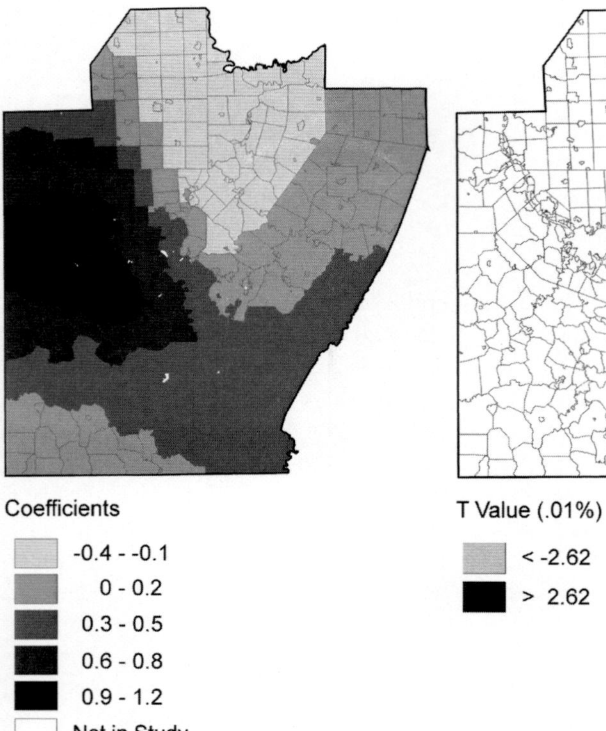

Fig. 4.6 GWR estimate of effect of a Comprehensive Plan on change from non-forested open-space to high-density urban-built environment

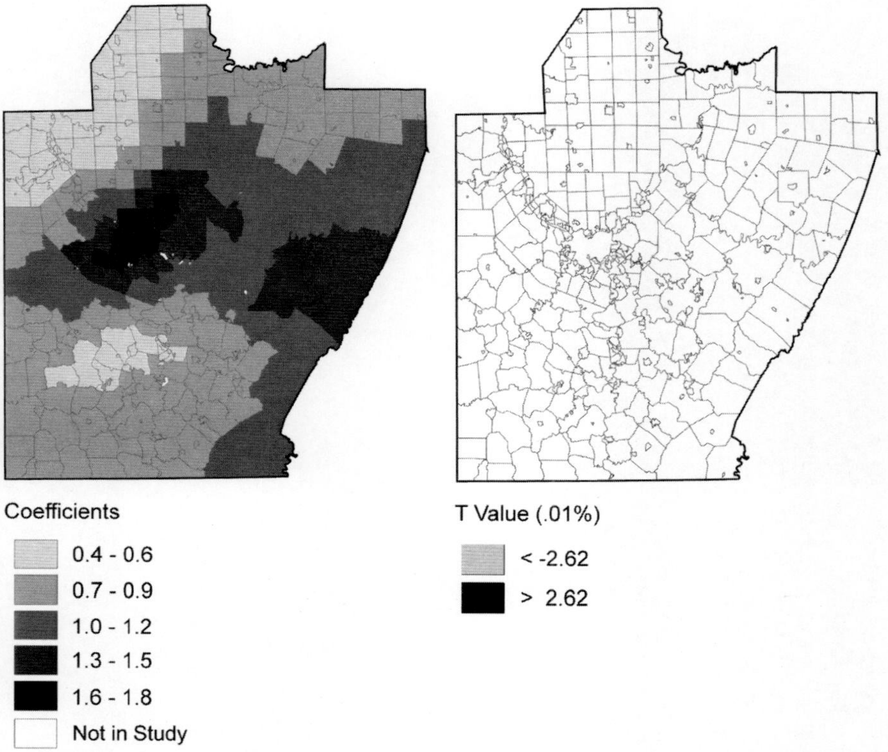

Fig. 4.7 GWR estimate of effect of a Zoning Ordinance on change from non-forested open-space to high-density urban-built environment

4.6 Conclusion

This research shows that the spatial fragmentation of local land-use policy has a variable impact on the change in land-use from non-forested open space to an urban built environment. In southwestern Pennsylvania, this is most evident in the change to a low density urban built environment, but less so in the change to a high density urban built environment.

Local land-use policy itself is only intended to manage local land-use, and not necessarily land-use change. Although the relationship between local land-use policy and land-use is well defined, it is unclear what the relationship should be, a priori, between local land-use policy and land-use change. It is clear, however, from the evidence of this research that local land-use policy does affect urban growth variably. Furthermore, it does so in different ways depending on the various planning tools administered by the local municipalities. The nature of the spatially non-stationary relationship between local land-use policy and urban land-use change was uncovered by GWR.

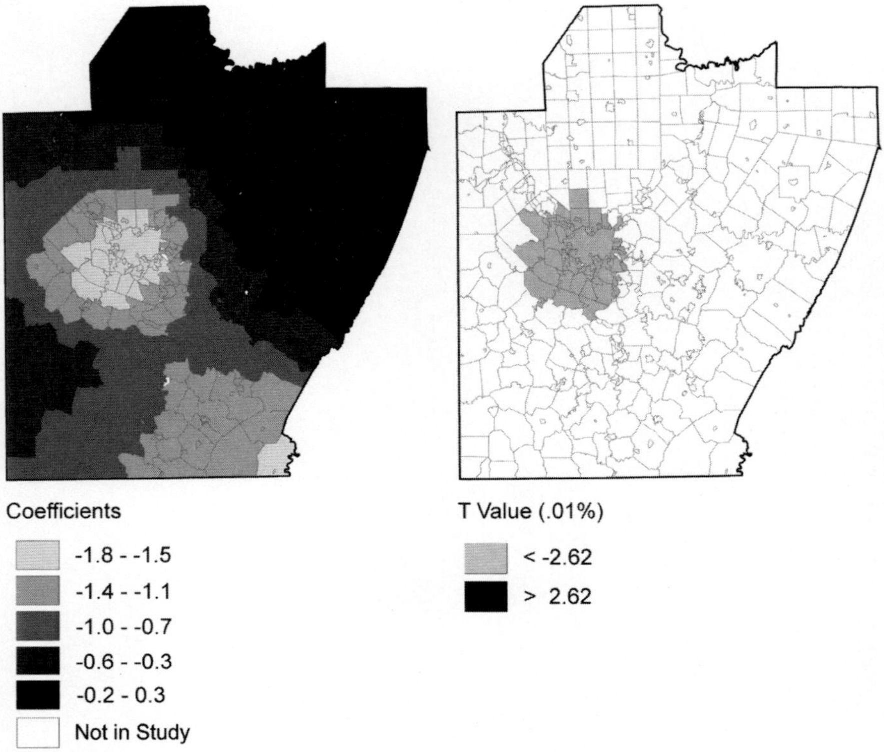

Fig. 4.8 GWR estimate of effect of a Subdivision Regulation on change from non-forested open-space to high-density urban-built environment

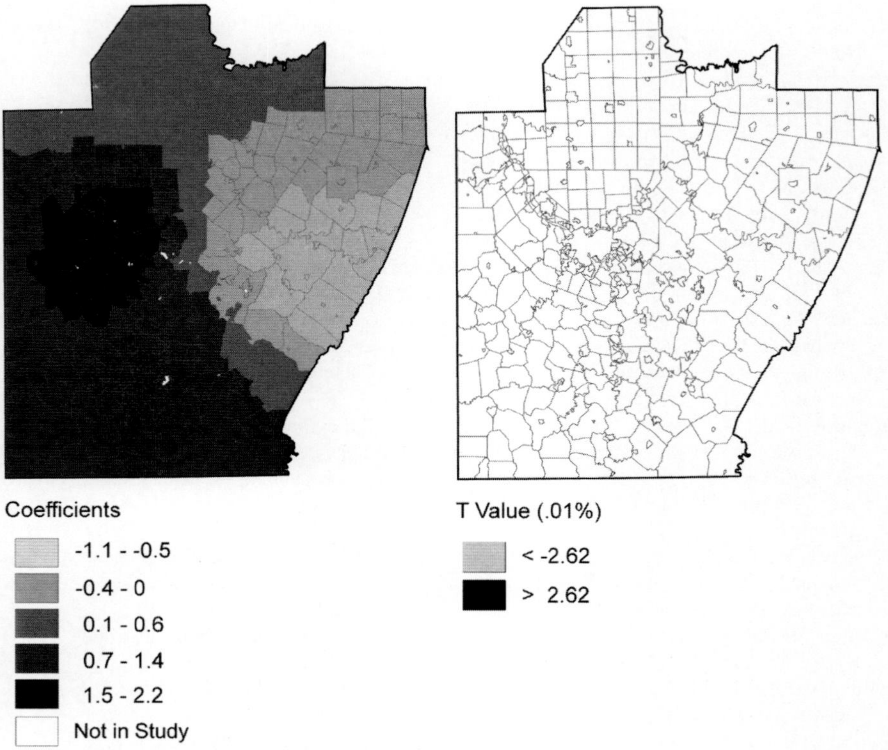

Fig. 4.9 GWR estimate of effect of a Planning Commission on change from non-forested open-space to high-density urban-built environment

References

Briem, C. n.d. *A Primer on Local Government Fragmentation and Regionalism in the Pittsburgh Region: Breakdown of local governments by MSA*. http://www.briem.com/frag/PittsburghIndex.htm. Compiled using U.S. Census Bureau, Census of Governments, 1997. Washington D.C.

The Brookings Institution Center on Urban and Metropolitan Policy. (2003). "Back to Prosperity: A Competitive Agenda for Renewing Pennsylvania." Retrieved from: http://www.brookings.org/es/urban/publications/pa.htm

Brueckner, J. and H. Kim. 2003. Urban Sprawl and the Property Tax. *International Tax and Public Finance*, 10: 5–23.

Brunsdon, C., A. Fotheringham, and M. Charlton. 1996. Geographically Weighted Regression: A Method for Exploring Spatial Nonstationarity. *Geographical Analysis*, 28: 281–98.

Burchell, R., C. Lowenstein, W. Dolphin, C. Galley, A. Downs, S. Seskin, K. Still, and T. Moore. 2002. Costs of Sprawl – 2000. Transportation Cooperative Research Program Report 74. National Academy Press, Washington, D.C.

Carruthers, J. 2002. Evaluating the Effectiveness of Regulatory Growth Management Programs: An Analytical Framework. *Journal of Planning Education and Research*, 21: 391–405.

Carruthers, J. 2003. Growth at the Fringe: The Influence of Political Fragmentation in the United States Metropolitan Areas. *Papers in Regional Science*, 82: 475–499.

Carruthers, J. and G. Ulfarsson. 2002. Fragmentation and Sprawl: Evidence from Interregional Analysis. *Growth and Change*, 28: 31–340.

Carruthers, J. and G. Ulfarsson. 2003. Urban Sprawl and the Cost of Public Services. *Environment and Planning B: Planning and Design*, 30(4): 503–522.

Charlton, M., A. Fotheringham, and C. Brunsdon. 2002. Geographically Weighted Regression Version 2.x Users Manual. University of Newcastle http://www.ncl.ac.uk/~ngeog/GWR/The_GWR_Manual.doc

Clawson, M. 1962. Urban Sprawl and Speculation in Suburban Land. *Land Economics*, 38(2): 99–111.

Daniels, T. 2001. Coordinating Opposite Approaches to Managing Urban Growth and Curbing Sprawl. *American Journal of Economics and Sociology*, 60: 229–243.

Fotheringham, A., C. Brunsdon, and M. Charlton. 2000. Quantitative Geography: Perspectives on Spatial Data Analysis. Sage, London.

Hoch, R. 2005. An Analysis of Fragmented Land-use Policy and Land-use Change: The Case Study of Metropolitan Pittsburgh. Unpublished Doctoral Dissertation. West Virginia University.

Lembeck, S. and T. Kelsey. 2001. How Effective Is Land Use Planning in Pennsylvania? No. 10 in the Land Use Planning in Pennsylvania Series, College of Agricultural Sciences, Penn State University.

Lembeck, S., T. Kelsey and G. Fasic. 2001. Measuring the Effectiveness of Comprehensive Planning and Land Use Regulations in Pennsylvania. The Center for Rural Pennsylvania. Pennsylvania General Assembly, Harrisburg, PA.

Lopez, R. and H. Hynes. 2003. Sprawl in the 1990s: Measurement, Distribution and Trends. *Urban Affairs Review*, 38: 325–355.

Mills, D. 1981. Growth, Speculation and Sprawl in a Monocentric City. *Journal of Urban Economics*, 10: 210–226.

Peiser, R. B. 1989. Density and Urban Sprawl, *Land Economics*, 65(3): 194–204.

Razin, E. and Rosentraub, M. 2000. Are Fragmentation and Sprawl Linked? North American Evidence. *Urban Affairs Review*, 35(6): 821–836.

Wassmer, R. W. 2006. The Influence of Local Urban Containment Policies and Statewide Growth Management on the Size of United States Urban Areas. *Journal of Regional Science*, 46(1): 25–65.

Chapter 5
GIS, Ecosystems and Urban Planning in Auckland, New Zealand: Technology, Processes and People

Fraser Morgan and Eric W. LaFary

Abstract This chapter discusses the roles of geographic information systems (GIS) in urban management, planning and ecosystem management within two regions of northern New Zealand. Situating the place of GIS is crucial towards elucidation of the context within which adoption and uptake has occurred. A review of the pertinent laws regarding local governance with emphasis upon planning and resource allocation will frame the chapter detailing the political space enveloping GIS within New Zealand broadly and the Auckland Regional Council (ARC) specifically. The structure of GIS as a department within the ARC follows. Three examples of GIS use, as a tool and increasingly as policy instruments, facilitating urban, rural and coastal management are related with an additional example of conceptual modeling described. The first study details the use of GIS as employed in urban residential land-use and rural landscapes through determination of building sections available for development. The chapter then examines the role of coastal modeling and its predictive capacities within resource management, urban planning and coastal protection. Finally, an exploration as to where GIS management and planning are traveling in New Zealand ensues by examining the role of exploratory dynamic modeling at Environment Waikato and Landcare Research whilst considering the potential for impact upon New Zealand urban and rural land management and environmental policy creation.

Keywords Ecosystems · Urban planning · Resource management · GIS · Auckland

5.1 Legal Landscapes: From Managing Resources to Understanding Ecosystems

Midway through the 20th century a number of Acts of Parliament relating to the structure of New Zealand's governance were passed. Namely, the Municipal

E.W. LaFary (✉)
The University of Auckland, School of Geography, Geology & Environmental Science, Auckland, New Zealand
e-mail: e.lafary@auckland.ac.nz

J.D. Gatrell, R.R. Jensen (eds.), *Planning and Socioeconomic Applications*, Geotechnologies and the Environment 1, DOI 10.1007/978-1-4020-9642-6_5,

Corporations Act and the Counties Act of 1954 and 1956, respectively, determined the manner in which councils were to govern. These acts defined the extent and roles that local and county governments were required to and restricted from performing in regards to the citizenry each determined by the central government. In essence, the Acts prescribed the social services and infrastructure that municipalities were to provide for communities. While powers were prescribed, local governments interpreted laws in a disparate manner and with little regard to environmental effects. The next major overhaul to local New Zealand governance passed parliament in 1974 as the Local Government Act 1974 (LGA 1974). LGA 1974 restructured the political landscape consolidating numerous governing authorities and councils. The prevailing governmentality experienced a metamorphosis from a nationally centered kaleidoscope of innumerable political entities entailing prescribed powers to a system of governance based upon a set of malleable guiding principles of management. While the streamlining effects of the LGA 1974 were somewhat effectual, the system of governance remained fragmented as authorities and councils interpreted the various laws regarding service provisioning and resource allocations in particular, in sometimes widely varying manners impacting the well-being of citizens and environment alike.

In 1991 the Resource Management Act (RMA 1991) was passed into law repealing nearly 70 previous articles of legislation. Notably, the Water and Soil Conservation and Town and County Planning Acts were subsumed by the new law. The enactment of this legislation radically altered the manner in which authorities and councils managed communities and interacted with the bio-physical components of the national landscape. The RMA 1991, by incorporating all facets of resource use under a single document, insists that its underlying principle is adhered to in all developmental and resource consents. While the LGA 1974 effectively outlines the legal responsibilities set out for the nation's authorities, territories and regions, the RMA is the defining law dictating the sustainable use of nearly all air, water, mineral and coastal resources. While, the notion of sustainability as the bedrock for environmental policy is seemingly ambiguous today, with the passing of the RMA 1991 New Zealand was one of the first nation-states to adopt such a far reaching stand on balancing cultural, social, economic and environmental well-being; not to mention the inclusion of a bio-physical baseline to ensure sustainable management. As a resultant consequence, the concept of sustainability and the practice of sustainable management were set as the cornerstones for New Zealand's governmentality moving into the 21st century.

The Local Government Act 2002 (LGA 2002) was a response, in part, to the needs of local authorities burdened with the pressures of governing the demands of rapidly changing population dynamics across the country that were no longer aligned with the dictates of the LGA 1974. Further, central government developed an awareness that local councilors, managers and mayors in combination with local citizenry were best able to identify and ensure delivery of goods and services to the country's disparate localities. The LGA 2002 restructured the role and scope of local and regional government outlining the governing boundaries for 73 local districts or territorial authorities located within 12 regional authorities. Each local body has a mayor and a council while the regional authorities have a CEO and managing council elected on 3-year cycles. The LGA 2002 also devolved most control of daily activities of governing to the local bodies: the final step in the central government relinquishing management of communities and communal resources. The result was reinforcement of the perspective that involving members of the community in consultative processes in respect to decisions surrounding resource allocation and services to be provided to the community as a central imperative. No longer were the days of a nationally elected politician determining the most efficient use of resources for communities hundreds of kilometers away. Local and Regional Authorities were charged with understanding the needs of citizens and ultimately provisioning resources to reflect those desires in a sustainable manner. The consultative process leads to community outcomes which are statements distilled from the full consultations. This new governmentality – local-sustainability with direct accountability to the general public – encountered many challenges for policy makers and planners in the Auckland Region. Largely due to population increases, the resulting strain upon infrastructure and resources were occurring at levels far greater than the rest of the country.

The Auckland Region, located in the upper North Island is south only of the Northland Region (Fig. 5.1). Coasts and oceans are dominant features on the

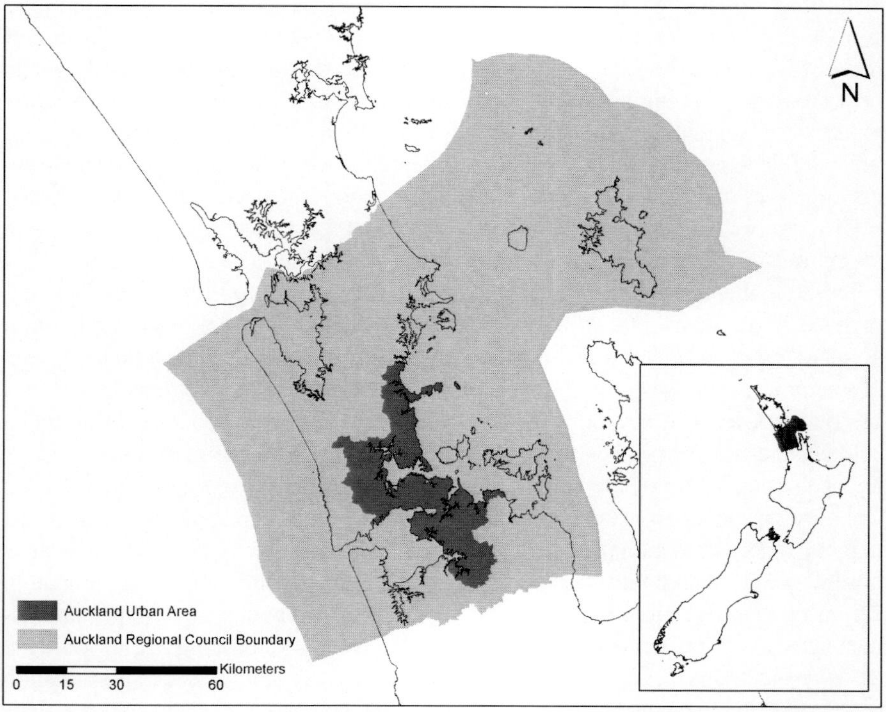

Fig. 5.1 Auckland Regional Council boundary and Auckland Urban Area (boundary data supplied by Statistics New Zealand)

Auckland landscape with the Tasman Sea, the Hauraki Gulf and the Waitemata and Manakau Harbors all providing physical regional boundaries. Unrelenting population increases over the past several decades have forced policymakers and planners to rethink the methods of decision making and policy construction employed in respect to Auckland's bio-physical resources. Charged with the practice of sustainable management whilst ensuring the cultural, economic, social and environmental well-being of the region, the Auckland Regional Council (ARC) recognized the need to adopt progressive means for resource management, infrastructural development and public transport planning. The reality of managing large quantities of information in combination with the demand to make sustainable decisions – many with repercussions upon the four well-beings far into the future – cemented the decision ARC took to be an early adopter of GIS as an integrated resource for planning and management.

5.1.1 Evolving GIS

The use of GIS within the ARC has evolved significantly over the past decade. Initially the organization used GIS as a specialized tool whose main purpose was the storage and management of cadastral and zoning information which was then analyzed for planning based activities. The relatively recent worldwide rise in the value of geospatial information is clearly reflected in the organizational structure of many of New Zealand's regional authorities with Auckland's regional body as the most progressive in implementation of GIS. One of the more visible changes in the uptake of geospatial data management within the ARC can be seen in the move to make highly specialized data and tasks available on desktops throughout the organization. This deployment of GIS services throughout the organization has resulted in GIS and related technologies being viewed as a fundamentally core resource in respect to urban planning and ecosystem management throughout the region.

In line with the small scale operation in the early 1990's, staff who were actively involved in the management and use of the cadastral data, were associated within the planning department of the organization. The change to a stand alone integrated GIS team has established the GIS unit as an essential, stand alone department similar to other aspects of the organization such as, environment, economy or heritage. This change has provided a number of opportunities to the organization through the steady expansion of the team and the centralization of core data and services. One of the more significant benefits to the ARC was the comprehensive web-based mapping and information platform that provides information to the public as well as more detailed information to service and policy staff throughout the organization. Currently, there are eight full-time staff members participating in the GIS management activities. The move to eight employees represents a doubling of staff over the last 10 years and is a clear indication as to the high level of importance placed upon geospatial knowledge in Auckland. The team consists of a team leader, two developers and five analysts. In addition, the team's close relationship with The University of Auckland provides direct access to students who are routinely employed for short periods for specific projects. Even with the occasional input from students, the team is unable to keep up with the recent demand that the various internal departments place upon them. Outside of the team's mandatory requirements (activities such as location based regional rates allocation, mapping for all aspects of the regional plans, and maintaining external and internal mapping applications) the demands for geospatial information throughout the Council sometimes result in a lag-time which occasionally reaches six months. Consequently there has recently been the need to engage external consulting contracts to keep up with the internal demand for geospatial work.

Historically, the GIS team had been utilized as a department that you sent your mapping needs to in order to have an image created for publication or for use within policy discussions. The evolution within the ARC's GIS team has revamped the environment from service provider to a necessary component of policy creation, as geospatial knowledge increases in value by managers, consultants and councilors. A key example of this forward thinking behavior is ARC's Capacities project which was undertaken to analyze the potential areas available to housing development in order to address the longstanding debate over the size of the ARC-controlled metropolitan urban limits. This and three other aspects will illumine the case studies presented below.

5.2 Urban Planning: City Boundaries and Coastal Ecosystems

A central component of ecosystem planning in New Zealand, the Auckland region in particular, is planning with a regional focus. The RMA 1991 mandated the sustainable use of all resources implicating governing authorities to adopt management regimes differing from traditional methods of strict local control of bio-physical resources. Namely, the move from specific resource allocation to a much more broad and complicated vision of ecosystem approach management was necessary to ensure the sustainable use of Auckland's cultural, social, economic and environmental values. Whilst, the benefits of having bio-physical bottom lines for effective management controls were largely greeted with acclaim, the difficulties inherent to implementation of an ecosystem strategy – effectively treating the entire region as one congruous space – were quickly realized as a limiting factor. The need to manage vast amounts of geospatial data, balancing not only static aspects of the physical environment but also more dynamic social and cultural attributes ensured the rapid adoption of GIS skills within the ARC.

5.2.1 Capacity for Growth: Managing Urban Sprawl

Historically, the Auckland region has seen residential development focused upon vacant land, with infill and business development less prevalent. Specifically 40% of residential development has been on vacant land, 32% as infill and 28% has occurred upon properties zoned for business activities (Gamble and Horsley 2008). Whilst Auckland has no shortage of undeveloped land for continued residential development, the impacts of an ever-expanding urban boundary impacts the ability of ecosystems to effectively function, ultimately altering the capacity for services valued by citizens to be delivered. Therefore, the governance structure in Auckland's regional authority, one based upon the principles of sustainability, recognized the need to re-think the prevailing development strategy. Traditionally, the ARC had maintained a supply of vacant land by extending the Metropolitan Urban Limit (MUL) to match growth, placing added strain upon the region's ecosystems with unknown negative externalities. The ARC was concerned with the reality that the supply of vacant land was being used up quickly and more importantly unsustainably as a result of an unprecedented period of population growth. While the original population estimates produced by the Regional Growth Strategy anticipated an increase of more than one-third in the next 50 years, recently updated population projections now suggest that this is more likely within the next three decades. With the ARC apprehensive about continually expanding the MUL to meet demand, a need existed to highlight and measure the areas where development could occur in line with the Regional Growth Strategy and within the MUL (Gamble and Horsley 2008).

5.2.1.1 Implementation of Growth Capacities

The 'Capacity for Growth' study has been developed by the ARC to meet these requirements through monitoring and reporting on the residential, business and rural land availabilities within the region. The ARC's Regional Policy Statement requires that these surveys are undertaken once every 5 years (alongside the most recent 2008 survey, it was also completed in 2003 and 1998) for the explicit purposes of managing urban containment (ARC Regional Policy Statement 1999) and to monitor the progress and implementation of the Regional Growth Strategy (ARC Regional Policy Statement 1999). Using a custom GIS application, the study investigated three states of land which could be developed upon (Vacant, Brownfield and Residential). Each of these states were examined in detail to assess each parcel's potential for development. The findings of the report outlines the amount of residential and business land available in the region for future development and the likely years of demand this capacity will satisfy. The ability to monitor the progress of particular policy initiatives is a resultant consequence of the capabilities of GIS management. As a result of the findings for this project, ARC is able to construct policy that ensures dense urban development rather than continual expansion of the MUL and the resultant effects of unchecked sprawl. The role of GIS in urban planning in Auckland has grown so that policy is being constructed with geospatial knowledge as a central attribute towards a data management component.

Vacant land was identified through an ARC built GIS application which used Color Digital Aerial Photography (resolution of 0.125m per pixel), Digital Cadastral Database (parcel boundaries) and District Plan zoning supplied by the local territorial authorities. Both residential and business zoned vacant land was investigated. This application required human input to ascertain if the parcel was vacant or partially vacant, in this case the percentage of area which was undeveloped was also recorded. While business zoned vacant land was recorded in hectares, vacant residential land calculated the number of potential dwellings based upon the relevant local council's zoning density levels while also accounting for the provision of access, open space and utility reserves.

Brownfield land was included in the analysis in response to the reality that vacant land was not the sole source of land for business development. As Gamble and Horsley (2008) state, "the redevelopment of previously used sites within the urban area has been part of the city's on-going development and is a reflection of the dynamics operating between the age of the stock, changing business activities and practices, changing locational needs and changing land values amongst other things. Including this measure provides a more comprehensive assessment of the availability of business land for business growth." Using a specific set of criteria (business zoned land, parcel size greater than 5000 m^2 and have an improved value less than 30% of the capital value), geospatial cadastral databases were queried to present a number of candidate Brownfield parcels which could be redeveloped. These candidate parcels were then manually assessed using the custom GIS application on-screen using the aerial imagery based upon the level of on-site activity, condition of structures and buildings, and formality of use if the parcel was used for storage.

In regards to this chapter, the most relevant use of the custom built GIS application was to investigate the ability for each residential parcel to undertake a forum of infill development. Infill-development is the term given to the development of dwelling(s) to the front or rear of an existing residentially developed site. Widespread infill began in central Auckland suburbs and then rapidly spread through most pre-1980 suburbs (subdivisions post-1980 contained lot sizes in line with the minimum lot size currently allowed). The definition can be expanded to cover infill-(re)development where the site is completely redeveloped with the removal and replacement of the original house with two, three or more townhouses. Both infill-development and infill (re)development are analyzed within the GIS application.

The first step in the custom GIS application was to source and load all the necessary databases into the application. These databases included district plan zoning (from all seven local councils), zoning density assumptions (checked with local councils), zoning development controls (yards set-backs, access way requirements, bush protection, etc), parcel attributes (Digital Cadastral Database), dwelling data (sourced from Quotable Value New Zealand), building footprints, and aerial imagery (Gamble and Horsley 2008). From these databases an initial round of parcels was identified by querying each parcel using the GIS databases and identifying those parcels that would theoretically meet the requirements for residential infill development and infill-redevelopment.

To calculate infill development potential, the resulting parcels were then individually assessed on screen using the GIS application. For parcels with digital building footprints available the GIS application would make an assessment of the availability of a dwelling(s) and its associated access-way. In these cases the assessment involved confirming or amending the on-screen data. In cases where digital footprints were not available the first step of the assessment involved digitizing the existing dwelling outline. The GIS application then calculated the building platform and access-way areas. The calculations were then confirmed or amended based upon the user's input.

The calculation of infill-redevelopment (the removal of the existing dwelling and redevelopment to the maximum permitted density), was performed automatically by the custom GIS application on all of the parcels that met the theoretical requirements for residential infill (Fig. 5.2). Two measures of infill-redevelopment were recorded: a purely theoretical level and modified realistic level. The theoretical measure simply captured the capacity on all parcels where the maximum number of potential dwellings is greater than the existing number. The modified count only records capacity where the maximum number of potential dwellings is at least twice the existing number (i.e. used as a proxy for sufficient incentive to redevelop) (Gamble and Horsley 2008).

Advancements in geospatial technologies have allowed policymakers to write policy that is more effective in regards to the principles of sustainability. The ability of planning and policy personnel to develop initiatives based upon alternative future scenarios predicated upon grounded data is proving to be a great strength when attempts are made toward environmental management. The application of GIS

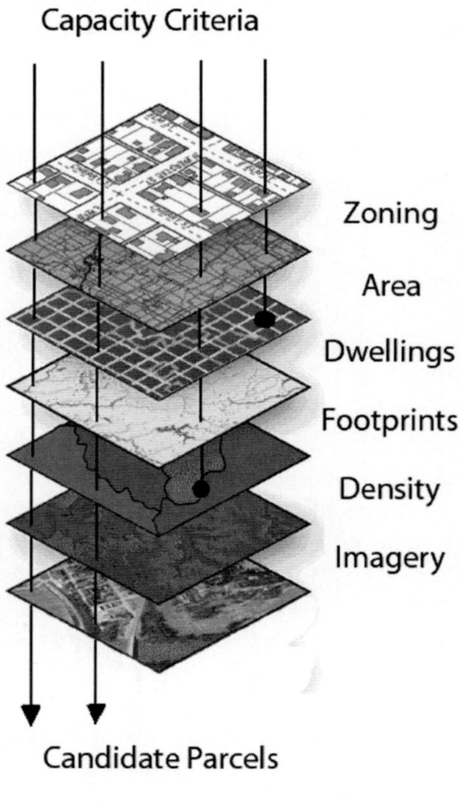

Fig. 5.2 Calculating parcels which meet the requirements of residential infill (from Gamble and Horsley 2008)

based knowledge has been the foundation for this forward-looking outcome based planning and management.

5.2.2 Estuaries, Beaches and Society: Modeling and Managing Auckland's Coasts

The environmental management of the coastal sphere is one of the core responsibilities of New Zealand's regional councils. The ARC is one of the more proactive regional authorities in this area, developing geospatial solutions both internally and externally to assist in the management, maintenance and protection of the coastal ecosystem. Consequently, the use of GIS in managing the coastal environment has altered the way in which the ARC views the coastal environment and the environmental issues encountered. The employment of GIS to manage the coastal environment has enabled the ARC to discover that the common concept of assessment of environmental effects is only one component of the decision making process (Green and Hatton 2007). By using additional tools, such as spatial predictive modeling, the ARC has been able to take into account a wide range of issues, values and opportunities whilst optimizing development opportunities and minimizing risks to the marine environment. Three levels of use within the ARC follow the aspects of management, maintenance and protection of the coastal ecosystem.

5.2.2.1 Coastal/Estuarine Modeling

The first example, protection, is aligned within the core operations of the GIS team, through the creation and distribution of numerous coastal policy maps. A substantial number of maps are produced regularly which outline the current coastal policy to be enforced. The second, maintenance, is one of the key areas of interest for the ARC. Understanding how the coastal environment is under pressure from problems which have their respective origins on the land is crucial to maintaining a pristine coastal environment. As Green and Hatton (2007) discuss, erosion of disturbed soils associated with land development and the generation of heavy-metals in the urban landscape are two of the main issues to face the coastal environment. Solutions to these problems are also land based – sediment-retention ponds, better stormwater treatment, and control of heavy-metals at the source – and are central to effective coastal management. ARC's ability to understand the effects that the different physical domains (land, streams, estuary, and coastline) have on each other is solved through spatial composite modeling.

The Okura estuary (Fig. 5.3), located in the northern part of Auckland, is under continued pressure from urban development. This pressure raised questions from the local and regional government about carrying capacity of the ecosystems, in terms of residential development, for the currently rural catchment. National Institute of Water and Atmospheric Research (NIWA) worked with the ARC to create a detailed scenario model which gave the regional and local councils the ability

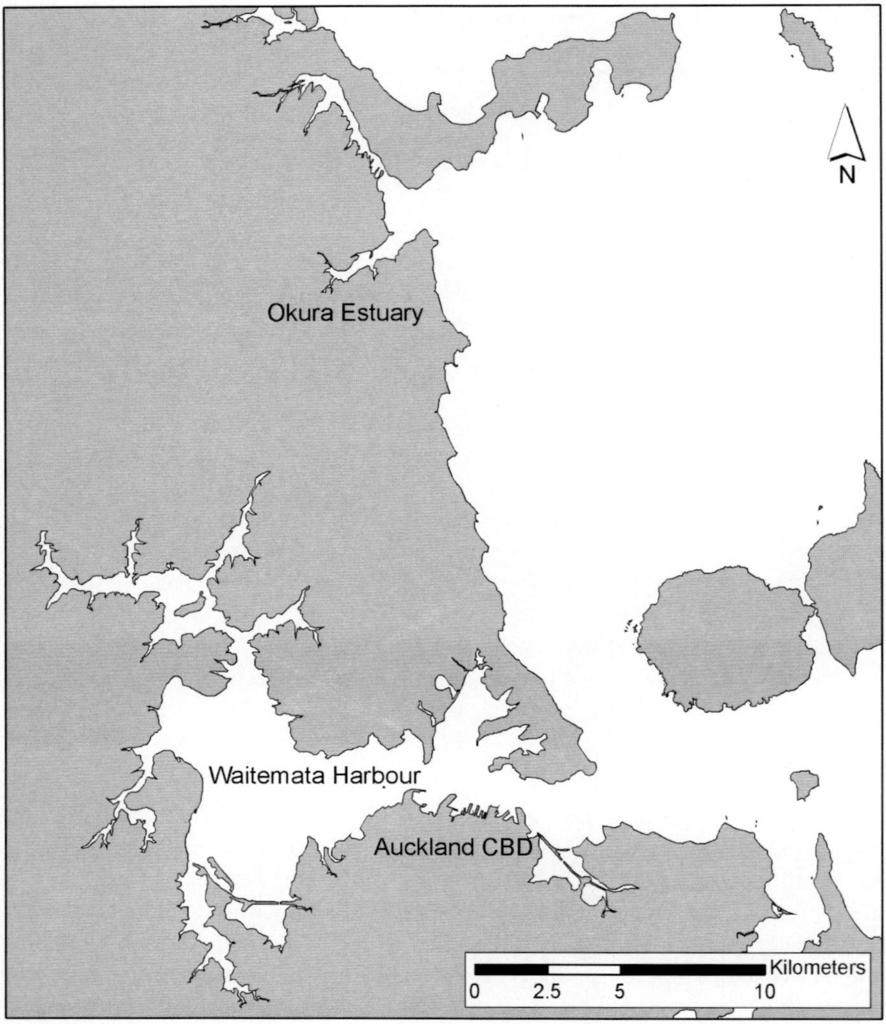

Fig. 5.3 Detailed view of the Auckland Region, showing the Okura Estuary and Waitemata Harbour

to understand the impacts of rural intensification. The second model, focused on Auckland's Waitemata Harbour (Fig. 5.3) investigates how existing development surrounding the harbor has increased the generation of heavy-metals flowing into waters from land-based activities (Green and Hatton 2007). These heavy-metals have a wide effect, from public health issues through to altering the balance of nutrients for the local marine ecosystem. The results of the model depicted the balance between environmental consequences and informed the ARC's choices about how the area should be developed and managed. Ultimately, policy is constructed with these findings alongside potential future scenarios as visualized through a GIS platform.

The use of geospatial technologies by the ARC to manage the coastal environment can be found in their development of the potential areas of aquaculture within the region (Pardy et al. 2003). In response to significant pressure on coastal space for aquaculture, a moratorium was placed on new locations of aquaculture to enable regional councils to develop Aquaculture Management Areas (AMAs). In developing a policy framework to manage aquaculture development, the first step was to define areas which should be managed. Taking into account the spatial extent of the various competing uses and values, areas that could be compromised by aquaculture or could constrain successful aquaculture activities were identified. This mapping was undertaken through a combination of user input, ARC's spatial databases and ESRI desktop GIS software. All of the information layers collected or created for the development of the AMAs continue to be used for other purposes such as the contaminate modeling discussed above.

The benefits of GIS and spatial information within the ARC, in regards to the coastal environment, are particularly noteworthy. Most geospatial information which is analyzed and displayed within GIS requires 'hard' lines to be drawn around the spatial extent of the interested aspect. In the case of the ARC, the practice of drawing representative boundaries around numerous subjective constraints and the need to visually represent commonly held concepts often helped to crystallize their thoughts and positions. As Pardy et al. (2003) states "sometimes this process revealed the apparent disagreement over verbal descriptions largely evaporated once they were translated into visible spatial areas of constraint".

The use of GIS and in particular the visualization of dynamic modeling within the framework is central to ecosystem management in the Auckland Region. GIS has enabled policy makers and citizens to visualize the spatial extents of protected regions and areas of restricted human activity. Whilst removing the 'fuzzy' aspects of boundaries never drawn and often reinterpreted, widely available maps and model output also create new scenarios for managers and planners to contend with. Namely, the opportunity for citizens to access highly detailed information in regard to boundaries. As data becomes publicly distributed interested citizens become armchair GIS technicians and planners interpreting data without adequate training and understanding of the myriad of policies and restrictions associated with the now neatly defined boundaries of admitted activity. Finding ways to manage data and answering public enquiry in creative ways is now on the agenda of policy makers and managers relying heavily upon widely available GIS data.

5.3 What Next? Conceptual GIS and Regional Planning

The use of GIS in ecosystem management and urban planning is central to the capabilities of regional managers and inseparable to the process of policy construction in Auckland. However, as models have become more reliable and the original dictates of GIS as policy and management tool are realized, the push for more attribute data resulting in models whose output evolves from a static planning tool to predictive instrument. Linking behavioral characteristics of New Zealand's populace to static data sets has the potential to radically alter how policy is written and the manner in which regions are managed. Two examples of conceptual models and the implications in regard to planning and ecosystem management are illustrated in the following conceptual studies around Regional Futures, and behavioral aspects of urban development. Development of dynamic models levies potential for governance to move from effects-based management of ecosystems to proactive planning to sustainably manage the environment. These conceptual models are filling the niche realized in order to plan for all dynamics at play within the regional sphere, ranging from coastal and freshwater processes through to individual human behaviors. The general consensus amongst policymakers and regional managers is that it is better to act and move towards a more sustainable regional environment, rather than not acting at all.

5.3.1 Choosing Regional Futures

One conceptual tool that is currently being developed for use for regional governance in respect to ecosystem management in New Zealand is the 'Creating Futures' project (www.creatingfutures.org.nz). The 4-year government-funded research project, which began in July 2006, is aimed at developing, testing and implementing integrated tools designed to inform communities of the long-term effects of current development patterns and trends and to enhance choosing and planning for desired futures. Initially, the project is being applied to the Waikato region, a central North Island location that covers approximately 25,000 km^2 (Fig. 5.4). The region is extremely diverse, with central volcanic mountains, high country plateau, the country's longest river (Waikato River) and largest lake (Lake Taupo), coastal plains and rugged coastline (Fig. 5.4). Much of the Waikato region has been converted to primary production with 55% of the region supporting agriculture particularly dairy/beef/sheep farming (McDonald and Patterson 2003). Further, the region contains the country's fourth largest city, Hamilton, which has evolved from an agricultural service town to a large, vibrant city.

With the focus of the project being regional in scope, the regional council, Environment Waikato (EW), leads the interdisciplinary team of social, environmental and economic researchers within New Zealand (Landcare Research – LCR, AgResearch – AG, New Zealand Centre for Ecological Economics – NZCEE, National Institute for Water and Atmospheric Research – NIWA, Scion, University of Waikato – UoW) and internationally (Research Institute for Knowledge Systems (RIKS) – Netherlands and the Centre for Economics and Ethics of the Environment and Development (C3ED) – France) The project has two objectives, firstly to develop processes to enable evaluation, deliberation and choice of alternative futures for social, environmental, economic and cultural changes through the use of scenario analysis linked to multi-criteria evaluation frameworks. The second, is to develop a spatial decision support system (SDSS) that integrates key aspects of the economy, environment, and society/culture (in a manner similar to Engelen et al. 2003; Sengupta and Bennett 2003; Rutledge et al. 2008). The SDSS will allow

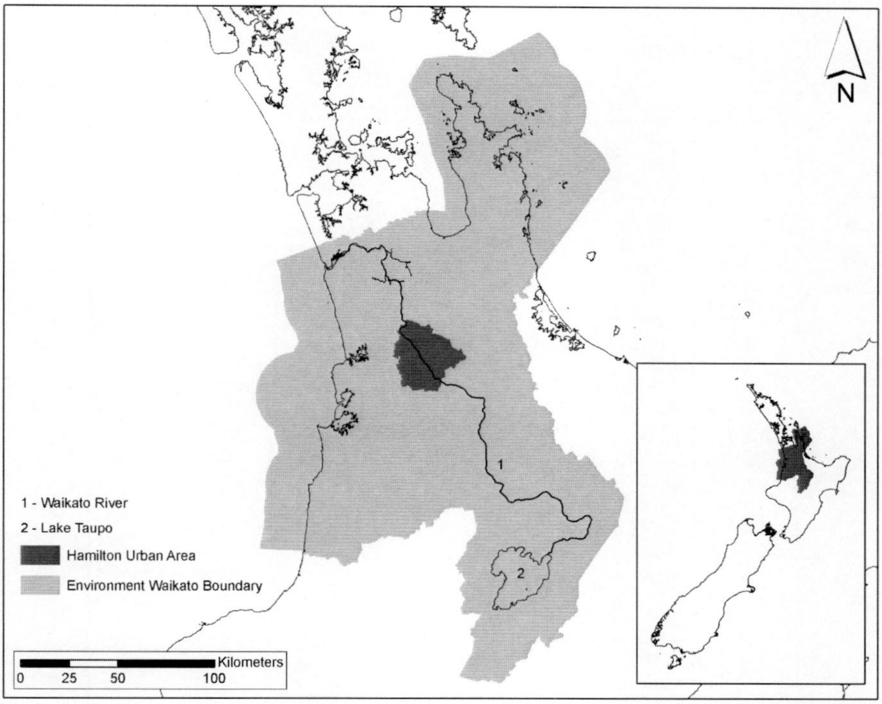

Fig. 5.4 Environment Waikato Regional Council and Hamilton Urban Area Boundaries (boundary data supplied by Statistics New Zealand)

users to explore plausible futures of regional development in a quantitative and spatially explicit manner, evaluate and compare different policy and planning strategy options, and help monitor and report on progress towards achieving long-term sustainable community goals and outcomes.

5.3.1.1 Modeling Future Scenarios

The overall system design of the Choosing Regional Futures SDSS consists of a series of integrated model components that operate at four spatial scales: NZ & the World, regional, district, and local (i.e. 200 m × 200 m grid cells) (Rutledge et al. 2007). The overall system diagram (Fig. 5.5) outlines the plan for the SDSS. Each model component represents more detailed sub-models, each of which vary in detail, data, and internal structure. Arrows represent links or flows among the different model components. The direction of the arrow shows the direction of flow. For example, the Demography model component generates information on population at the district level. Population is pooled among all districts to create a supply of labor to the Labor Market, which in turn supplies demand for workers from the Regional Economy. The SDSS is being developed and run under Geonamica®, an object-oriented application framework developed by RIKS to build decision support

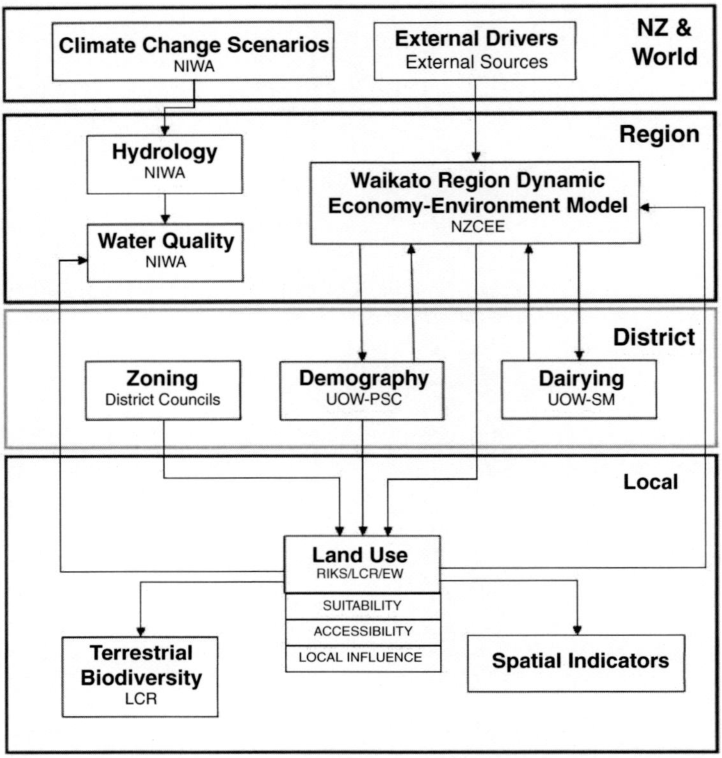

Fig. 5.5 Choosing Regional Futures SDSS diagram (from Rutledge et al. 2008)

systems using spatial modeling and geosimulation (Hurkens et al. 2008). It has been developed over the past 15 years and has been used to generate numerous integrated spatial decision support systems in Europe.

At the national and global scale, Climate Scenarios consist of pre-defined scenarios for climate change. The scenarios represent standard IPCC climate change scenarios that have been down-scaled for use at scales relevant to New Zealand (Scenarios based on the average output from 12 Global Climate Models compiled for the IPCCs Fourth Assessment Report – http://www.ipcc.ch/ipccreports/ar4-syr.htm). The External Drivers contain information on key drivers that may strongly influence New Zealand and/or the Waikato region. These variables could include factors such as foreign exchange rates, world commodity prices, interest rates, credit availability, migration trends, technological developments, etc.

The regional scale examines the economy along with hydrologic and water quality information. The economy is modeled using a system dynamics model of region-wide environment-economy interactions (McDonald 2007). It models the flow of economic commodities as traded on markets in both monetary and physical terms, along with the flow of associated natural resource inputs (e.g., land, energy, water) and residual outputs (e.g., wastes, pollutants and emissions). Hydrology is simulated using a purpose-built model of surface and shallow groundwater hydrology to generate annual estimates of runoff. Water Quality is modeled through an adaptation of the SPARROW (http://water.usgs.gov/nawqa/sparrow/) model (Elliott et al. 2005). SPARROW estimates pollution loads, such as total nitrogen or total phosphorous, at a given point in the river network as a function of upstream catchment characteristics including land use.

At the district scale, Zoning, Demographics, and Dairying models are accounted for. Zoning consists of zoning maps prepared by district and city councils as part of their district/city plans. Zoning information directly affects the local-scale land use model by controlling where different land uses can and cannot occur. Demographics are modeled through the University of Waikato's Population Studies Centre (http://www.waikato.ac.nz/wfass/populationstudiescentre/) (Cameron et al. 2007). The model estimates birth rates, death rates, and net migration rates (in male and female 1-year increments) from each district to other districts within the region, and from each district to outside the region. The Dairying component models different levels of dairy intensity within each cell based on a combination of physical attributes or suitability, production targets (i.e. kg/milk solids/ha desired), and management practices.

Locally, Land Use and Biodiversity are modeled. The Land Use model component dynamically models land use change over time based on demand for certain land uses generated at the district (Demographic) or regional (Economic – i.e. demand for land) scales and a combination of four other factors: Zoning (see above), suitability, accessibility, and local influence (Hurkens, Hahn, and van Delden 2008). Suitability estimates the biophysical suitability of land for different uses. For example, areas with steep slopes will not be suitable for certain types of agriculture. Accessibility typically relates to travel distances, which can affect the desirability of some land uses such as housing. Local influence measures the influence of neighboring land uses on a land use at a particular point. The four factors combined with regional demands change land use potential over time, resulting in land use changes. The Biodiversity component tracks changes in indigenous and exotic vegetation over time. It combines information on vegetation with information on protected areas and the Land Environment of New Zealand terrestrial environmental classification (http://www.landcareresearch.co.nz/databases/LENZ/) to provide information on biodiversity status across a range of scales (local, district, and region). Changes in Land Use could affect Biodiversity both positively (restoration) or negatively (removal) (Rutledge et al. 2004; Walker et al. 2005).

The project has been designed to ensure close interaction between researchers and end-users so that the resulting SDSS can be integrated into strategic planning and policy making and resource management by regional and local councils. The question of exactly how the SDSS will be used by councils is an active component of the research. By linking researchers and end-users from the beginning, the project serves as a direct pathway for uptake of the information, tools and knowledge gained by councils and allows them to gradually build their capacity and capability using tools like the SDSS to inform integrated, long-term planning for the future.

5.3.2 Behavioral Considerations in Planning: Theoretical Modeling Within NetLogo

The urban development process, through which the built environment is produced and used, is critical to the understanding of urban development and local and regional governments' attempts at managing the complex systems which result from it. At the heart of the process of urban development are residential real estate developers who drive and motivate urban development investment decisions (Healey and Barrett 1990). While their role is well understood in a general sense, research into their types, strategies and behaviors remains limited. Most empirical analyses treat developers as an undifferentiated whole, whose sole objective is to maximize their profit (Coiacetto 2000; Guy and Henneberry 2002). This profit maximizer approach has been adopted within existing spatial models of the urban development process (Healey 1991). In reality, developers are a diverse group varying in their size, financial resources, and assessment of the market (Coiacetto 2001). As a consequence, a number of non-rational behaviors arise (such as territoriality, cooperation, and risk avoidance) which moves the developer away from a pure profit maximizer position. These non-rational behaviors play a major role in the way they react to the social, environmental and policy issues which surround the land space they aim to develop. Ignoring these subtle, and not so subtle, differences leads to confusion and conflict between both sides – leading to increased costs, delays and defensive behavior. Staff at Landcare Research is currently developing a model which will enable an understanding of the developers' motives, behaviors and responses and to help regional and local government to be aware of the socio-economic constraints that their policy places upon developers.

5.3.2.1 Model Development

The development of behavior and social interactions between the developers and councils is created through a spatial agent-based framework. An initial version of the model is being developed within the NetLogo modeling environment (Wilensky 1999, http://ccl.northwestern.edu/netlogo/). While NetLogo has recently introduced the ability to integrate GIS information into the modeling environment, the initial model is based upon an abstract representation of a 'city'. Using a binary tree data structure, the abstract 'city' is created inside a cellular world. While a fully developed world will account for in excess of 65,000 individual lots, the initial state of the model creates a 'city' with around 6000 lots, which represent a realistic cross-section of lot sizes. Transportation networks are not included within the model as it is assumed that the networks will adapt to meet the growth of the city. Within the abstract 'city', a three level market is included to account for areas of greater property value. The three layers (global, neighborhood, local) contribute to a combined property value on a scale of 0 to 1. While global is fixed, neighborhood and local value can change after each round of developer behavior based upon the level of development which occurs within the 'city'. Based upon a fixed level of capital for the model, a number of property developers are created which have a variety of behaviors. Within the model, behaviors such as territoriality and risk management are associated to the level of capital each developer begins with.

Currently, the model is ready to examine how behavior can change the way a city develops. Each developer within the model is given an opportunity to either purchase and/or develop a parcel of land. The developer evaluates the percentage of available parcels against three distinct aspects; the existing available capital the developer has access to, the quality of parcels currently available, and the developer's assumption of the market's future. If the developers' purchase criteria are met, the developer will purchase the parcel. Each developer then chooses whether a parcel they own should be developed. This parcel could have just been purchased, or parcel purchased in a previous round. This decision is one of the more crucial, as it is the timing of these decisions which ultimately affect the form and growth of the city. In making this decision, each parcel the developer owns is assessed to find the parcel which at the time will produce the best return. It is important to note that the developer can choose not to subdivide any parcel because they believe the market will improve. This choice is based upon the current value of the parcel, the ongoing costs of holding onto the parcel for another round, and the developers' assessment of the market. These factors can be distilled into a function to equate as to whether the parcel should be held for another cycle. If the decision to subdivide a parcel is made, the final decision the developer makes is how intensive the development should be. This decision takes into account the level of development surrounding the parcel, the current property value of the parcel, and the financial situation of the developer. The interface of the model (Fig. 5.6) showcases and records each developer decision for further analysis and presents a visual record of how the 'city' grows.

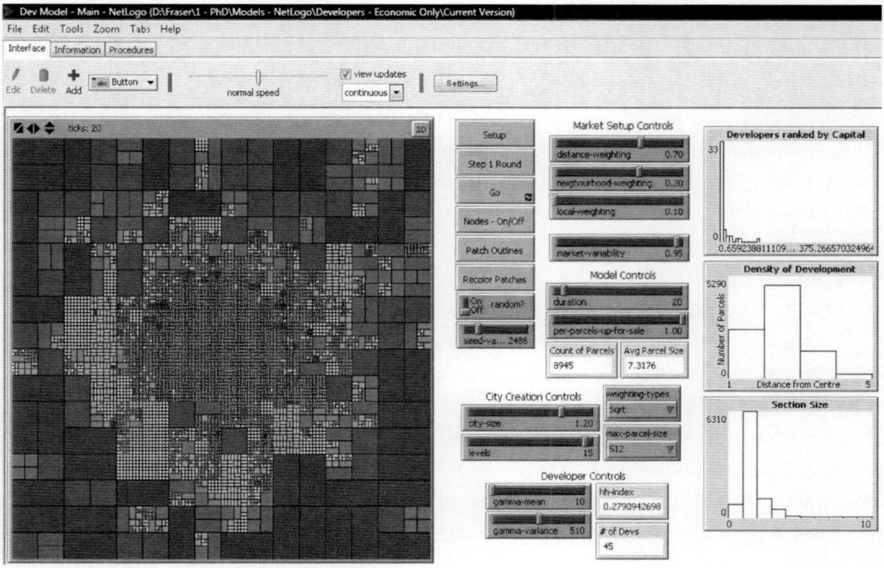

Fig. 5.6 Screenshot of the Developer Behaviour Model

Whilst development on the model is ongoing, initial impressions are that the addition of certain behaviors does have an effect on the way in which a city is developed. While the current 'abstract' model has not been applied to real world data, the final model has been designed to be scale and geographically independent as to provide access to these tools at a wide range of scenarios and locations. It has also been designed to be used as a module to fit within other wider scale models of urban growth and development. Being aware of how residential real estate developers would respond to policy change would provide a shift in behavior from both the end-user (higher quality policy, which will accurately shape the growth of the city) as well as the developers (less conflict and greater acceptance of the policy), both of which would contribute to the achievement of more desirable urban design outcomes through more focused urban development investment.

5.4 Concluding Remarks

The integration of GIS from novel mapping tool to an integrated policy instrument has experienced a rapid evolution. New Zealand began implementing GIS as map making activity in the 1980s. The swift increase in technological ability allowed planners and managers to begin to utilize GIS as a means of looking forward rather than a static representation of spaces, places and boundaries connected to areas of protection and policy. As New Zealand's governance structures began to reflect naturally bounded ecosystems the country's legislation began to reveal the ability to write law as intended for implementation through a GIS interface. The place of GIS within policy creation and ecosystem management grew from an ancillary characteristic of regional and local bodies to fully integrated into virtually all aspects of looking after and planning ahead for the cultural, social, economic and environmental well-beings of the nation. Whilst the integration of GIS platforms and output has allowed policymakers and planners to be forward thinking, the availability of data and the accuracy of maps also provide new challenges for governing entities to confront. Citizens can now access high resolution, low scale maps where private homes can be identified allowing individuals from communities to ascertain what laws and environmental mandates apply to their properties – an unintended consequence of policy written for GIS application that is currently being resolved in New Zealand. Nonetheless, the use of GIS for regional planning and environmental protection has allowed managers to be more effective as data are available on-screen, anytime.

Charged with the sustainable management of all resources, regional councils have increasingly turned to the capabilities of future outcomes planning – by means of dynamic modeling and more traditional GIS visualization. Detailed data has been compiled for all lands that are developable in Auckland to ensure that the region has sustainable housing potential to accommodate increasing immigration pressures and also to combat difficulties witnessed by sprawling cities throughout the world. The continual increase of citizens in the region also place considerable pressure upon coastal ecosystems' capacity to provide goods and services necessary and desired for the sustainable management of communities. Findings from coastal and ecosystems modeling efforts are utilized as baselines for ecosystem function and are consequently fundamental to ecosystem management. Environment Waikato's Choosing Regional Futures initiative is an example of a comprehensive attack on controlling destructive development and related activities of population increase and demands on physical resources. Choosing Regional Futures is the future of ecosystem management and urban planning in New Zealand as numerous attributes are entered into the interface to visualize alternative futures. GIS, in this instance, is more than a mapping tool, as policy may be scripted from the model's output as the region can be managed from a digital perspective. Dynamic spatial models are being developed in New Zealand to account not only for environmental and built realities, but also to enter human behavior into the predictive outcomes. The promise for planners and managers to picture the effects of actors driving urban expansion and by default stress upon ecosystems is quickly being realized through research undertaken across New Zealand.

The confluence of national scale environmental policy initiatives and technological advances created an atmosphere in New Zealand placing the traditional management perspective of planners and policy against a strict GIS outlook. However, as the advantages toward data management, predictive modeling and visualization of the landscape within a GIS were realized the demand for geospatial knowledge amongst planners and policymakers rose dramatically. The outcome has been one of more informed management based upon the principles of sustainability and regional governance increasingly demanding GIS tailored ecosystem policy.

Acknowledgments We would like to thank the following individuals and organizations for the many contributions toward this chapter. Vivienne Cole, Shaun Gamble and Dominic McCarthy, of the Auckland Regional Council, for consultation and assistance with acquiring reports and images. Daniel Rutledge from Landcare Research for assisting with information relating to the Regional Futures project and Mal Green from NIWA for consultation on coastal modeling. The Choosing Regional Futures Team from EW, NIWA, Landcare Research, University of Waikato, NZCEES and EW for creating a collegial atmosphere during the construction of this chapter. Kathryn Hayward for reviewing and editing the manuscript. Finally, we wish to thank Jay Gatrell and Ryan Jensen for the opportunity to participate in this book project. Any errors in the text are oversights of the authors.

Abbreviations

AG	AgResearch
AMAs	Aquaculture Management Areas
ARC	Auckland Regional Council
CEO	Chief Executive Officer
EW	Environment Waikato
GIS	Geographic Information Systems
LENZ	Land Environment of New Zealand
LCR	Landcare Research
LGA 1974	Local Government Act 1974

LGA 2002	Local Government Act 2002
MUL	Metropolitan Urban Limit
NIWA	National Institute of Water & Atmospheric Research
NZCEE	New Zealand Centre for Ecological Economics
RIKS	Research Institute for Knowledge Systems
RMA 1991	Resource Management Act 1991
SDSS	Spatial Decision Support System
SPARROW	SPAtially Referenced Regressions On Watershed Attributes
UoW	University of Waikato

References

Auckland Regional Council (1999). Auckland Regional Policy Statement. http://www.arc.govt.nz/albany/fms/main/Documents/Plans/Regional%20Policy%20and%20Plans/ARPS/ARPS%20Policy.pdf. Auckland, New Zealand.

Auckland Regional Council (1999). Auckland Regional Growth Strategy: 2050. http://www.arc.govt.nz/albany/fms/main/Documents/Auckland/Aucklands%20growth/Auckland%20regional%20growth%20strategy.pdf. Auckland, New Zealand.

Cameron, M., Cochrane, W., Poot, J. (2007). End-user Informed Demographic Projections for Hamilton up to 2041. PSC Research Report. Hamilton: Population Studies Centre, University of Waikato.

Climate Change 2007 (2007). Synthesis Report. http://www.ipcc.ch/ipccreports/ar4-syr.htm. Intergovernmental Panel on Climate Change. Sweden.

Coiacetto, E. J. (2000). Places shape place shapers? Real estate developers' outlooks concerning community, planning and development differ between places. *Planning Practice and Research*, 15 (4), 353–374.

Coiacetto, E. J. (2001). Diversity in real estate developer behavior: A case for re-search developer strategies. *Urban Policy and Research*, 19 (1), 43–59.

Elliott, S.H., Alexander, R.B., Schwarz, G.E., Shankar, U. Sukias, J.P.S. and McBride, G.B. (2005). Estimation of nutrient sources and transport for New Zealand using the hybrid mechanistic-statistical model SPARROW. *Journal of Hydrology*, 44 (1), 1–27.

Engelen, G., White, R. and de Nijs, T. (2003). Environment Explorer: Spatial Sup-port System for the Integrated Assessment of Socio-Economic and Environmental Policies in the Netherlands. *Integrated Assessment*, 4 (2), 97–105.

Environment Waikato (2008) Choosing Regional Futures Project. www.choosingfutures.org.nz. Hamilton, New Zealand.

Gamble, S. and Horsley, S. (2008). *Capacity for Growth Study, 2008 Interim Re-port, Technical Publication 369*. Auckland: Auckland Regional Council.

Green, M. O. and Hatton, C. (2007). Predictions for use in Resource Management. In N. Holmes (ed.), *Reviews in Environmental Science and Management*. Lismore, Australia: Southern Cross University Press.

Guy, S. and Henneberry, J. (2002). *Development and Developers: Perspectives on Property*. Oxford: Blackwell Science.

Healey, P. and Barrett, S. M. (1990). Structure and agency in land and property development processes: some ideas for research. *Urban Studies*, 27 (1), 89–104.

Healey, P. (1991). Models of the development process: a review. *Journal of Property Research*, 219–238.

Hurkens, J., Hahn, B. and van Delden, H. (2008). Using the GEONAMICA® soft-ware environment for integrated dynamic spatial modelling. In *iEMSs 2008: International Congress on Environmental Modelling and Software*. Barcelona, Spain: International Environmental Modelling and Software Society (iEMSs).

Landcare Research-Manaaki Whenua (2008). Land Environment of New Zealand terrestrial environmental classification http://www.landcareresearch.co.nz/databases/LENZ/. Auckland, New Zealand.

McDonald, G. and Patterson, M. (2003). Ecological Footprints of New Zealand and its Regions. Ministry for the Environment. Wellington, New Zealand. http://www.mfe.govt.nz/publications/ser/eco-footprint-sep03/eco-footprint-sep03.pdf

McDonald, G. (2007). Auckland Region Dynamic Economic-Environment Model, Technical Report. Takapuna: Market Economics.

Pardy, S., McCarthy, D. and Moore, A. (2003). *The Development of Aquaculture Management Areas: A Spatial Analysis Approach.* Paper read at Coasts and Ports Australasian Conference, at Auckland, New Zealand.

Rutledge, D., Price, R., Heke, H. and Ausseil, A. (2004). National analysis of biodiversity protection status: methods and summary results. Landcare Research Contract Report LC0405/038 prepared for the Ministry for the Environment. Hamilton: Landcare Research

Rutledge, D., McDonald, G., Cameron, M., McBride, G., Poot, J., Scrimgeour, F., Price, R., Phyn, D. and van Delden, H. (2007). Choosing Regional Futures – Spatial Decision Support System. Draft Specifications. Hamilton: Landcare Research.

Rutledge, D., Cameron, M., Elliott, S., Hurkens, J., McDonald, G., McBride, G., Phyn, D., Poot, J., Price, R., Scrimgeour, F., van Delden, H., Tait, A. and Woods, R. (2008). Choosing Regional Futures – Spatial Decision Support System. Final Specifications. Hamilton: Landcare Research.

Sengupta, R. R. and Bennett, D. A. (2003). Agent-based modelling environment for spatial decision support. *International Journal of Geographical Information Science*, 17 (2), 157–180.

Walker, S., Price, R., Briggs, C. and Rutledge, D. (2005). Remaining indigenous covers on pastoral leases: priorities for biodiversity protection. Landcare Research Contract Report LC0405/163 prepared for Land Information New Zealand. Hamilton: Landcare Research

Wilensky, U. (1999). NetLogo. http://ccl.northwestern.edu/netlogo. Center for Connected Learning and Computer-Based Modeling. Northwestern University, Evanston, IL.

Chapter 6
Hyperspectral Applications in Urban Geography

Vijay Lulla

Abstract This chapter examines the scale and scope of hyperspectral remote sensing applications and presents a brief case study. As the case study demonstrates, hyperspectral approaches expand the range and accuracy of very fine scale urban studies.

Keywords Hyperspectral data · Remote sensing · Terre Haute

6.1 Introduction

Remote sensing developed as a technology that could be used to study features and phenomena that covered large geographical extents. To this extent remote sensing has been very effectively and successfully used in environmental studies, urban studies (Chou et al. 2005; Hardin and Jensen 2005; R. Jensen 2002; Li and Weng 2007; Mundia and Aniya 2005), agricultural and forestry studies (R. R. Jensen and Binford 2004; Lulla 2005) to name a few applications of remotely sensed data. However with advances in hardware and software technology newer sensors are being built that provide better spatial and spectral resolutions. These increased resolutions provide more detailed spectral and spatial data that may allow for very fine urban studies.

This chapter broadly outlines the current research trends in the field of remote sensing and describes how policy implementers can use the data and the results obtained from these trends. This chapter consists of (a) a brief discussion of remote sensing (b) a short explanation of multispectral and hyperspectral remote sensing (c) details of hyperspectral remote sensing and (d) a brief case study of hyperspectral remote sensing.

V. Lulla (✉)
Department Geography, Geology, and Anthropology, Indiana State University, Terre Haute, IN, USA
e-mail: vlulla@indstate.edu

J.D. Gatrell, R.R. Jensen (eds.), *Planning and Socioeconomic Applications*, Geotechnologies and the Environment 1, DOI 10.1007/978-1-4020-9642-6_6,

6.2 Remote Sensing

Remote sensing refers to the process of collecting data of a feature without being in physical contact with the feature which for all passive remote sensing is either the electromagnetic radiation emitted by the sun or other emitted energy. Sensors capture the electromagnetic reflectance characteristics of the features in both the visible and the invisible portion of the spectrum and store these reflectance responses which are used for analysis. Remote sensing can be a very effective tool to study features that cover large geographic areas and have very distinctive reflective characteristics. Hence, some of the most common applications of remote sensing technology are found in land use land cover classification, change detection (J. R. Jensen 2000; Liu and Lathrop 2002; Lu et al. 2004; Muttitanon and Tripathi 2005), vegetation health and inventory measurement, urban sprawl and deforestation.

All remote sensing data are expressed in terms of their various resolutions which express the quality of remote sensing data that is being used. These resolutions are: (i) spatial resolution (ii) spectral resolution (iii) radiometric resolution (iv) temporal resolution. Spatial resolution describes the smallest spatial unit that the sensor can distinguish correctly and is usually termed as either 'pixel size' or 'instantaneous field of view'. The general rule is that a sensor must have a spatial resolution of at least one-half the smallest dimension of the things being studied. Spatial resolution is analogous to scale on any conventional map. Radiometric resolution refers to the ability of the sensor to distinguish between very fine spectral differences, and temporal resolution describes how often a sensor re-visits or is able to re-image a particular area. Spectral resolution is of more importance to us which will be dealt with in more detail.

6.2.1 Multispectral Remote Sensing

The process of acquiring electromagnetic reflectance in multiple bands of the spectrum is called *multispectral remote sensing*. A description of different technologies used by different multispectral scanners to collect these data can be found in (J. R. Jensen 2000) and (Lillesand et al. 2004). Landsat Multispectral Scanner (MSS) and Thematic Mapper (TM) series sensors, France's SPOT (Satellite Pour l'Observation de la Terre), and the Indian Remote Sensing satellite (IRS) series are just some of the many different multispectral satellites that are used for collecting remote sensing data. Table 6.1 shows the spectral bands of Landsat ETM+, the latest in the Landsat series of satellites. More details of the Landsat program can be found at http://landsat.gsfc.nasa.gov/.

Despite the large bandwidths used by the multispectral scanners, the data collected by these sensors have proved useful in forestry, land use land cover classification (Anderson et al. 1976) and change detection. Various vegetation indices such as normalized difference vegetation index (NDVI), soil adjusted vegetation index (SAVI) (Huete 1988), which are obtained by the ratioing of infrared,

Table 6.1 Spectral range of all the bands in Landsat ETM+ (http://landsathandbook.gsfc.nasa.gov/handbook/handbook_htmls/chapter3/htmls/filters.html)

Spectral band	Spectral range (μm)	Bandwidth (nm)
1	0.450–0.515	65
2	0.525–0.605	80
3	0.630–0.690	60
4	0.775–0.900	125
5	1.550–1.750	200
6	10.400–12.500	2100
7	2.090–2.350	260

and visible bands have proved very useful in vegetation related studies. Research of urban areas such as quality of life measurement, urban heat island effect and, urban sprawl has been primarily conducted using the mid-infrared and the thermal infrared bands.

6.2.2 Hyperspectral Remote Sensing

Hyperspectral remote sensing refers to a system which records reflectance information in very narrow spectral ranges as contiguous bands across the electromagnetic spectrum. Hence, we obtain a much more complete reflectance profile of features. The advantage of having a spectral profile over many small spectral widths proves useful in features with complex spectral characteristics. A good example of this would be a succession forest in the Amazonian rainforest which is a complex ecosystem owing to the many levels of secondary succession present in has many levels of succession with only very subtle differences in spectral responses.

Hyperspectral data are generally collected by installing the sensor onboard an aircraft (e.g., fixed-wing airplane or helicopter). This provides flexibility as to when the data can be collected and allows us to control the spatial resolution of the data by controlling the altitude of the flight. Some examples of hyperspectral sensors are NASA's Airborne Visible/Infrared Imaging Spectrometer (AVIRIS), Compact Airborne Spectrographic Imager (CASI), and the Airborne Imaging Sensor (AIS). Since these sensors are flown onboard an aircraft the data collection is not dictated by the climate conditions and the revisit time issues that multispectral sensors pre- sent. Some of these sensors also provide the ability to select user-defined spectral bands for data collection. However, in 2000, NASA launched Earth Observing Satellite I (EO-I) which carried Hyperion, a satellite based hyperspectral sensor. Hyperion is capable of recording spectral information in 220 spectral bands, 0.4–2.5 μm of spectral range at 30 m ground resolution. Hyperion data are useful in applications relating to geology, mining, forestry, and many others.

Despite the aforementioned advantages of hyperspectral data, some issues need to be addressed for these data to be practically usable. Hyperspectral data with its numerous spectral bands need considerable resources in terms of data storage, analysis, display, and archiving. Most research conducted using hyperspectral data is carried out on a much smaller spectral subset of the available bands. This is achieved by post-processing the acquired hyperspectral data. The primary reason for selecting smaller spectral subsets is to reduce the data dimensionality and extract the most useful information from the continuous spectrum rather than using all the available information. Different data reduction techniques such as principal component analysis (PCA) and best band selection are some of the commonly used techniques. The criteria for selecting best bands are governed by the feature/process that is being observed. Best bands selection is determined by the feature that is being observed and its reflectance characteristics. Reflectance characteristics may then be compared with many publically available spectral libraries. Spectral libraries are maintained by organizations such as U.S Geological Survey (USGS) (Clark et al. 2007), JPL (http://speclib.jpl.nasa.gov/document.htm) and Johns Hopkins University (http://speclib.jpl.nasa.gov/documents/jhu_desc.htm).

6.3 Applications of Hyperspectral Remote Sensing in Urban Environments

For many remote sensing studies in urban environments higher spatial resolution is of much more importance than spectral resolution (J. R. Jensen 2000). Urban environments consist of broadly described built-up features that construct very complex environments. This complexity warrants the need for higher spatial resolution. A detailed description of which spatial resolutions would be best for different classifications using the USGS classification scheme can be found in (Anderson et al. 1976). Despite the necessity for high spatial resolution data, hyperspectral data can be useful for studies pertaining to urban forestry. Multispectral data with moderately high spatial resolution (15–30 m) have been successfully used for urban forestry studies (Hardin and Jensen 2005; R. R Jensen et al. 2003). With the availability of improved sensors with higher spatial and spectral resolutions such as AISA+ (spatial resolution: generally 1.5 to 5m and spectral resolution: 248 spectral bands for 0.4–1.1 μm) there should be more data available for urban environments. A case study examining the use of AISA+ sensor data ensues.

6.4 Case Study: Hyperspectral Remote Sensing of the Urban Forest in Terre Haute, Indiana, USA

Recently, a study using the AISA+ hyperspectral sensor was conducted in the urban area of Terre Haute, Indiana. The AISA+ sensor is capable of collecting 248 spectral bands data between 0.45 and 1.1 μm range of the spectrum with a spatial resolution of 1.5–5 m depending on the flight altitude. The AISA+ sensor is fully programmable and can be installed in virtually any aircraft with a standard aerial

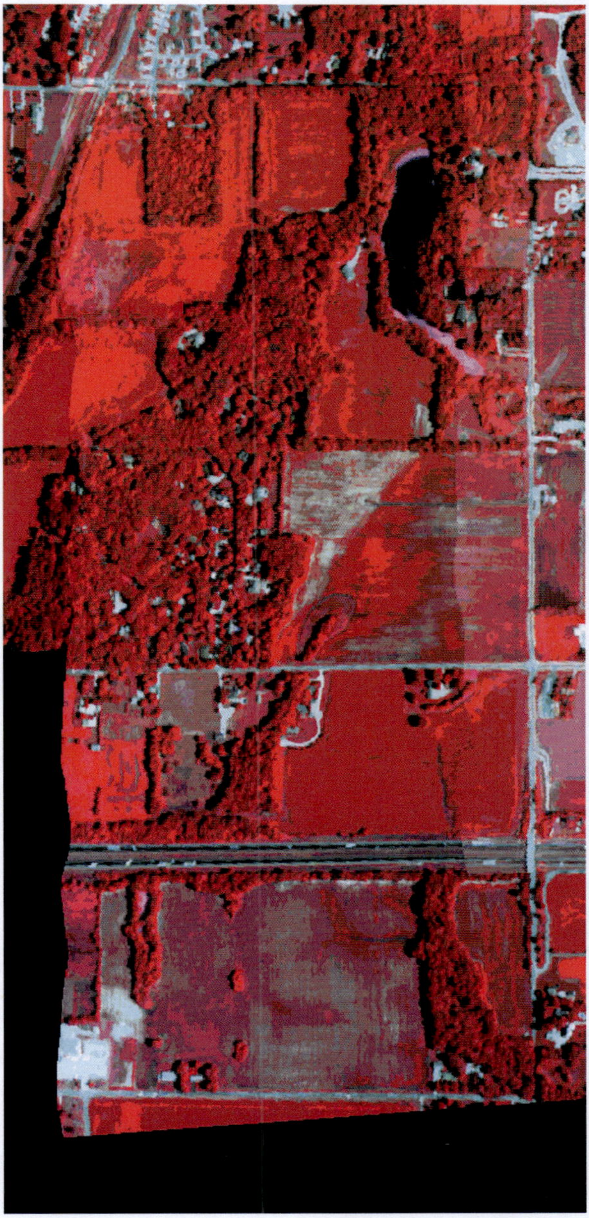

Fig. 6.1 Part of the 2 m resolution AISA+ data collected on 24, July 2006. Bands 174 (788 nm), 125 (673 nm), 84(577 nm) (R, G, B)

camera floor port. Figure 6.1 shows part of the data that was collected during this study. Spectral data collected by Landsat TM and AISA+ sensor for a cluster of trees from the below image are shown in Figs. 6.2 and 6.3 respectively.

Data acquired from the AISA+ sensor were used to estimate biophysical properties of the urban forest in Terre Haute, Indiana (Jensen et al. in press). These

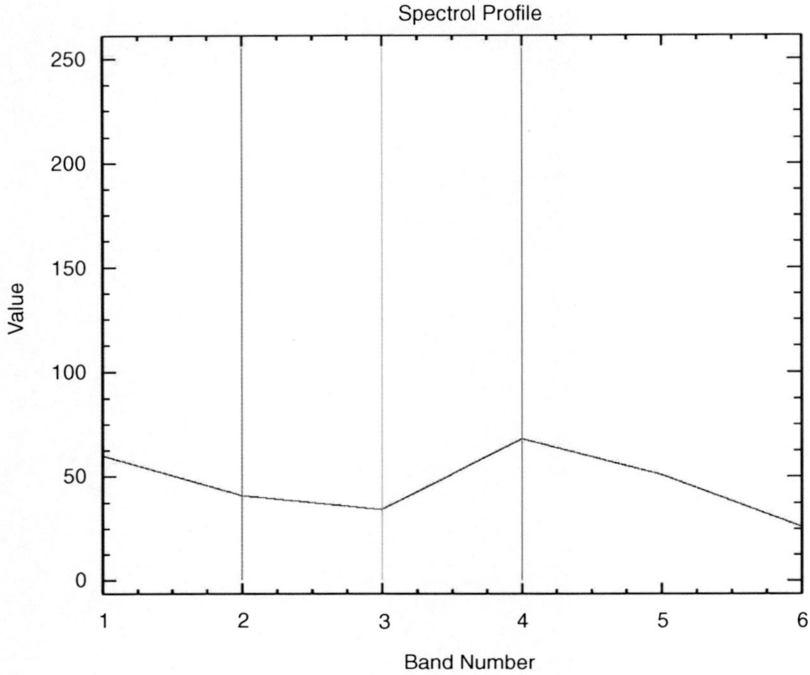

Fig. 6.2 Spectral response of tree cluster from TM data. Bands 4, 3, 2 (R, G, B). Note how the curve is under-sampled as compared to Fig. 6.3

data were correlated with in situ leaf area index (LAI) measurements of the urban forest and other urban environs throughout the city of Terre Haute. The authors used a variety of independent variables (e.g., individual radiometric bands, radiometric ratios) as input into stepwise regression models. They found that the regression coefficient of determination (R-squared) values ranging from 0.27 to 0.73. These results indicate that up to 73% of the variation in urban LAI can be explained by radiometric information measured by the AISA+ sensor. Radiometric features appearing most frequently in the models included band radiance at 727, 753, 848, 870, 900 and 917 nm. The best single predictor of urban LAI was the absolute difference in radiance between 777 and 673 nm.

6.5 Concluding Remarks

The use of hyperspectral data has been made possible by advancement in sensor technology and hardware and software improvements, which provide the ability to collect and analyze more spectral remote sensing data than was previously possible. Though the chapter has only briefly introduced what remote sensing and hyperspectral remote sensing are, a substantial amount of work is being carried out in the

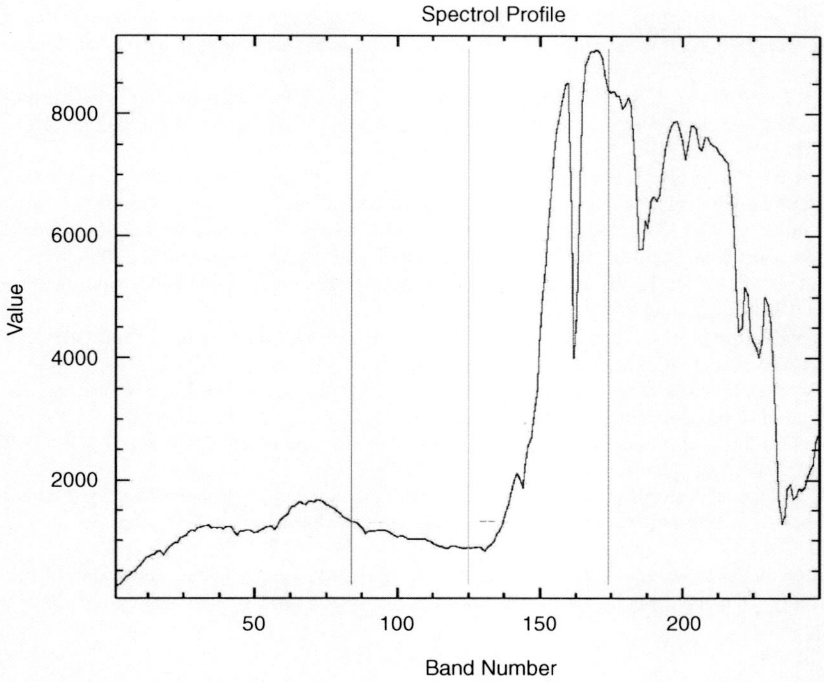

Fig. 6.3 Spectral response of same tree cluster from AISA+ data. Bands 174, 125, 84 (R, G, B). Note how the curve captures much more of the imaging spectroscopy

field to investigate novel and efficient ways of collecting, storing, processing and analyzing hyperspectral data.

References

Anderson, J. R., Hardy, E. E., Roach, J. T., and Witmer, R. E. (1976). A land use and land cover classification system for use with remote sensor data. Paper presented at the Geological Survey Professional Paper.

Chou, T. Y., Lei, T. C., Wan, S., and Yang, L. S. (2005). Spatial knowledge databases as applied to the detection of changes in urban land use. International Journal of Remote Sensing, 26(14), 3047–3068.

Clark, R. N., Swayze, G. A., Wise, R. A., Livo, K. E., Hoefen, T. M., Kokaly, R. F., et al. (2007). USGS Digital Spectral Library splib06a.

Hardin, P. J., and Jensen, R. R. (2005). Neural network estimation of urban leaf area index. GIScience and Remote Sensing, 42(3), 229–252.

Huete, A. R. (1988). A soil-adjusted vegetation index (SAVI). Remote sensing of environment, 25, 53–70.

Jensen, J. R. (2000). Remote sensing of the environment: an earth resource perspective. New Jersey: Prentice Hall.

Jensen, R. (2002). Spatial and temporal leaf area index dynamics in a north central Florida, USA Preserve. Geocarto International, 17(4), 47–54.

Jensen, R. R., and Binford, M. W. (2004). Measurement and comparison of Leaf Area Index estimators derived from satellite remote sensing techniques. International Journal of Remote Sensing, 25(20), 4251–4265.

Jensen, R. R., Boulton, J. R., and Harper, B. T. (2003). The relationship between urban leaf area index and household energy usage in Terre Haute, Indiana, U.S. Journal of Arboriculture, 29(4), 226–229.

Jensen, R.R., Hardin, P. J., Bekker, M. F., Farnes, D. A., Lulla, V., and Hardin, A. (in press). Modeling urban leaf area index with AISA+ hyperspectral data. Applied Geography.

Li, G., and Weng, Q. (2007). Measuring the quality of life in city of Indianapolis by integration of remote sensing and census data. International Journal of Remote Sensing, 28(2), 249–267.

Lillesand, T. M., Kiefer, R. W., and Chipman, J. W. (2004). Remote Sensing and Image Interpretation (Fifth edn): John Wiley and Sons Inc.

Liu, X., and Lathrop, R. G. (2002). Urban change detection based on an artificial neural network. International Journal of Remote Sensing, 23(12), 2513–2518.

Lu, D., Mausel, P., Brondízio, E., and Moran, E. (2004). Change detection techniques. International Journal of Remote Sensing, 25(12), 2365–2401.

Lulla, V. O. (2005). Biomass estimation using statistical and neural network analysis of ASTER data. Indiana State University, Terre Haute, IN-47809 USA.

Mundia, C. N., and Aniya, M. (2005). Analysis of land use/cover changes and urban expansion of Nairobi city using remote sensing and GIS. International Journal of Remote Sensing, 26(13), 2831–2849.

Muttitanon, W., and Tripathi, N. K. (2005). Land use/land cover changes in the coastal zone of Ban Don Bay, Thailand using Landsat 5 TM data. International Journal of Remote Sensing, 26(11), 2311–2323.

Chapter 7
GIS and Spatio-temporal Trends in Inequality: Tracking Profitability According to Firm Size in Japanese Manufacturing, 1985–2006

Shawn Banasick

Abstract This chapter examines the spatial patterns of profitability in Japanese manufacturing according to the size of the firm from 1985 to 2006. The study period captures the "bubble economy" of the late 1980s, the economic downturn and restructuring of the 1990s, and the economic expansion of the early 2000s. The Theil index of inequality and LISA cluster maps are used to identify trends in average profitability for six firm size categories. The results show that average profitability tends to decline along with firm size, and that instability in average profitability for larger size firms tended to be a major contribution to changes in total inequality. For much of the study period contributions from larger firms to inequality between the firm size categories was decreasing while their contributions to regional inequality were increasing. The analysis also suggests that there is substantial variation in average profitability trends for prefectures in the core manufacturing region.

Keywords Uneven development · Manufacturing · GIS · Japan

7.1 Introduction

In a capitalist economic system the spatio-temporal structure of capital accumulation undergoes continual transformation (Harvey 2006; Smith 1990). The regional restructuring of capitalist processes has attracted the attention of economic geographers for many years, but there have been relatively few studies that have provided a broad overview of the transformation of economic landscapes. An important exception is Smith and Dennis (1987) who examined the transformation of the American manufacturing core region from the 1940s to the 1980s. They focused on changes in the spatial clustering of manufacturing establishments, output and employment to argue that the restructuring of the economic landscape was more than simply a regional shift or the restructuring of relations of production in place, but entailed "a

S. Banasick (✉)
Department of Geography, Kent State University, Kent, OH, USA
e-mail: sbanasic@kent.edu

J.D. Gatrell, R.R. Jensen (eds.), *Planning and Socioeconomic Applications*, Geotechnologies and the Environment 1, DOI 10.1007/978-1-4020-9642-6_7,

complete restructuring of the scale at which regions are constituted as coherent and integrated economic units" (Smith and Dennis 1987, p. 171).

This chapter draws on the insights of Smith and Dennis' work to examine recent transformations in the Japanese industrial landscape. Japan experienced an extended period of economic decline during the 1990s after the collapse of a speculative bubble in real estate and the stock prices. The rising exchange value of the Yen coupled with the economic downturn led to intensified efforts by major manufacturers to internationalize their production systems as a means to reduce costs and maintain market access. Historically Japanese domestic production systems had been characterized by extensive networks of smaller sized subcontracting firms (Glasmeier and Sugiura 1991).

By looking at trends in profitability this chapter attempts to provide greater insight into the effects of the processes of restructuring according to the size of the manufacturing firm. As Harvey (2001) has argued, understanding changes in profitability is central to the examination of the processes of uneven capitalist development and therefore should provide the analysis with a better perspective on the processes of restructuring than an examination of just employment or output change.

In the first part of the chapter average levels of profitability for six classes of firm size are estimated on an annual basis for the 1985 to 2006 period. Next, trends in the profitability of the six firm size categories are examined through the use of a two-step Theil index of inequality. This approach tracks two dimensions of inequality in average profitability – between the firm size categories and between places for each firm size class. In the second part of the analysis the spatial patterns of the place-based contributions to inequality of each firm size class are examined through the use of the Local Moran statistic integrated into the ArcGIS software package. In the third part of the analysis the trends in profitability for firms in three prefectures of the core manufacturing region are examined in greater detail by ranking the performance of each firm size category relative to the profitability of the same category in the other forty-six prefectures.

7.2 Background

In the early postwar period Japan's manufacturing sector was often characterized as having a 'dual structure' in which small and medium firms were dominated by large firms. This perception was weakened by the dynamism that small firms demonstrated in the 1960s, and by the 1970s the extensive subcontracting networks of Japanese manufacturing were increasingly being perceived as providing a competitive advantage (Samuels and Keller 2003). The networks provided flexibility and efficiency gains and contributed to improved production practices as larger firms pushed smaller firms to improve the quality of their products (Kawai and Urata 2002; Kimura 2002; Whittaker 1997).

In the 1980s the economic climate for Japan's small and medium firms was dramatically transformed. In the first half of the decade the Japanese economy demonstrated relatively healthy performance despite a global economic downturn, with an average growth rate of approximately 3 percent (Harada and Onishi 2003). The G-5 meeting in 1985 proved to be a major turning point. At the meeting the US sought to revalue the dollar relative to other currencies, and the rapid strengthening of the Yen that resulted from these efforts dramatically undercut the relative competitiveness of Japanese manufacturers in international markets. When the Japanese government loosened monetary policy in an effort to ameliorate the effects of the stronger Yen it inadvertently helped to spark the emergence of a bubble economy. The bubble would last until mid-1990, and was centered on rapid price inflation of stocks and real estate (Dehesh and Pugh 1999). The Nikkei 225 Index peaked at a high of almost 40,000 at the end of 1989, but then as the bubble collapsed it dropped by almost 50 percent by mid-1990, and still only averaged 17,225 in 2006 (Hoshi and Kashyap 2004; Statistical Research and Training Institute 2008). Average prices for urban residential land also continued to decline from 1991 through 2007 (Statistical Research and Training Institute 2008).

The collapse of the bubble weighed heavily on the Japanese economy, and the 1990s were generally a period of poor economic performance that has been called Japan's 'Great Recession' (for example, Harada and Onishi 2003). Employment data suggest that smaller-sized manufacturing firms struggled to survive the economic downturn (Table 7.1). From 1985 to 2001 there were substantial declines in the number of full-time workers for all firm size categories, but the smaller firm size categories experienced the strongest declines. While some of these declines may be due to increased use of part-time and temporary workers, similar trends are evident in the number of establishments (Table 7.2).

Starting in 2002 Japan entered its longest period of economic expansion since the end of the Second World War (74 months by April 2008). Supported by strong growth in exports to Asia, the recovery helped push the 2005 operating rate in manufacturing to its highest level since the short-term recovery of 1997 (Economic and Social Research Institute 2008). The recovery also resulted in employment and

Table 7.1 Change in full-time workers by firm size

Firm size (# regular workers)	1985	2006	Percent change 1985–2001	Percent change 2002–2006
Over 300	3,076,686	2,447,563	−19.7	6.4
100 – 299	1,851,851	1,742,859	−6.1	4.7
30 – 99	2,237,268	1,743,331	−15.7	−3.1
20 – 29	1,046,467	736,428	−20.2	−5.6
10 – 19	1,167,937	824,131	−16.4	−10.5
4 – 9	1,509,740	731,130	−36.6	−15.0
All firm sizes	10,889,949	8,225,442	−18.6	−1.2

Note: Due to changes in Japan's industrial classification system, data after 2001 are not directly comparable to previous years.
Source: METI Research and Statistics Division.

Table 7.2 Change in establishments by firm size

Firm size (# establishments)	1985 Est.	2006 Est.	Percent change 1985–2001	Percent change 2002–2006
Over 300	3,764	3,371	−10.3	4.9
100–299	11,554	10,775	−6.5	4.1
30–99	42,308	32,220	−16.8	−3.8
20–29	42,791	30,036	−20.2	−5.7
10–19	84,506	60,515	−15.2	−10.6
4–9	253,595	121,626	−36.5	−15.7
All firm sizes	438,518	258,543	−27.9	−11.1

Note: Due to changes in Japan's industrial classification system, data after 2001 are not directly comparable to previous years.
Source: METI Research and Statistics Division.

establishment growth for the larger firms, but the smaller firms on average continued to lose full-time workers and establishments (Table 7.2).

7.3 Methods and Data

The first part of the analysis centers on estimating average profitability for Japan's six firm size categories. While Webber and Rigby (1996) have provided an extensive discussion on methods of estimating profitability in manufacturing, the data requirements for their approach prevents its application to small sized firms in Japan. An alternative procedure has been used by Schneider (1991) to estimate profitability for small manufacturing firms in Austria. For this estimation method profitability is calculated as surplus (value added – wages) divided by output (value added). In this fashion profitability was estimated for the six firm size categories for 46 of Japan's prefectures annually from 1985 to 2006 (Fig. 7.1). Okinawa prefecture was excluded from the analysis due to its very small manufacturing sector and its economic dependency on US military base-related income. The prefectures are designated as either core or periphery according to their share of national manufacturing employment in 1985.

The Theil index was then used to track trends in the levels of inequality in the estimations of average profitability for the six firm size categories. Recent attention has focused on adapting the Theil index to examine inequality over both time and space (see, for example, Galbraith and Garcilazo 2005; Galbraith et al. 2004; Akita 2003; Fujita and Hu 2001; Galbraith and Berner 2001). The Theil index is decomposed into two elements, one which measures inequality between places, and one which measures inequality between the firm size categories within places such that:

$$T = T_b + T_w, \tag{1}$$

where the between places (within firm size categories) Theil element is defined as

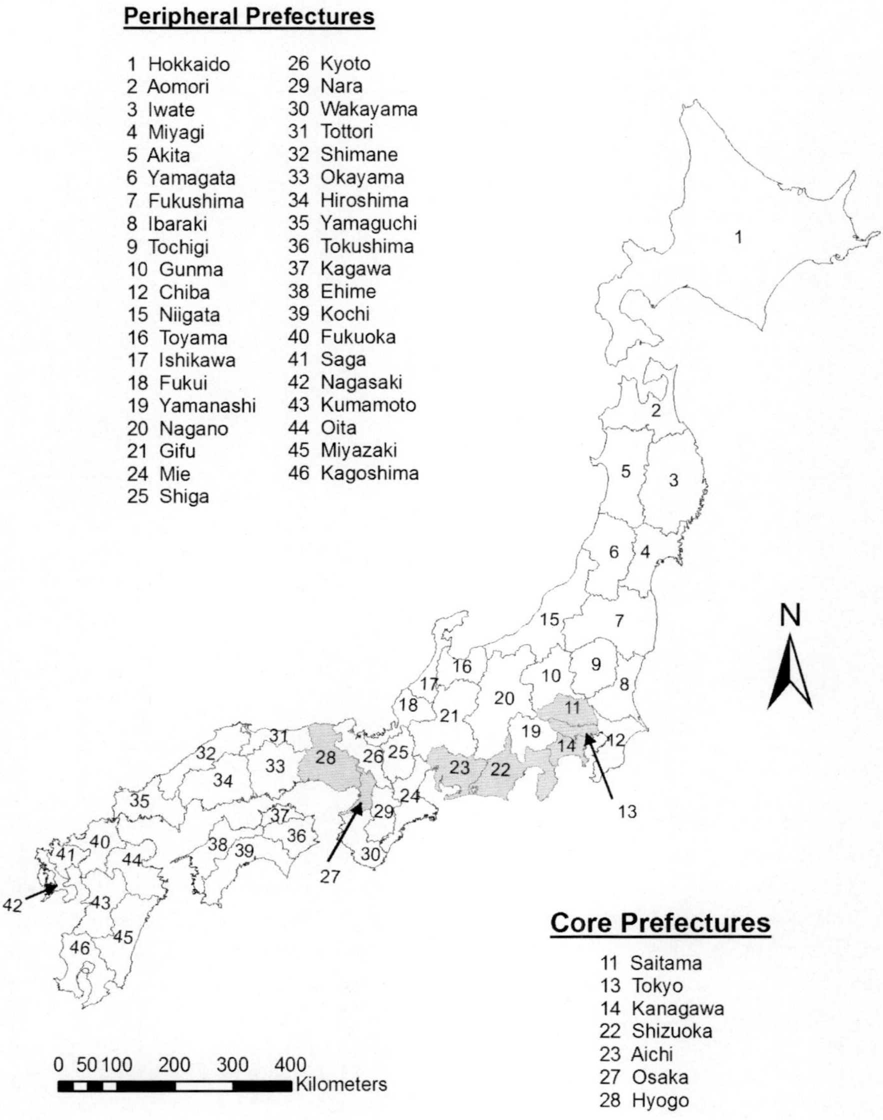

Fig. 7.1 Core and peripheral prefectures

$$T_b = \left\{ \sum_{i=1}^{n} \left(\frac{S_i}{\sum\limits_{i=1}^{n} S_i} \right) * \ln \left[\frac{S_i \Big/ \sum\limits_{i=1}^{n} S_i}{V_i \Big/ \sum\limits_{i=1}^{n} V} \right] \right\} \tag{2}$$

and the within place (between firm size categories) Theil element is defined as

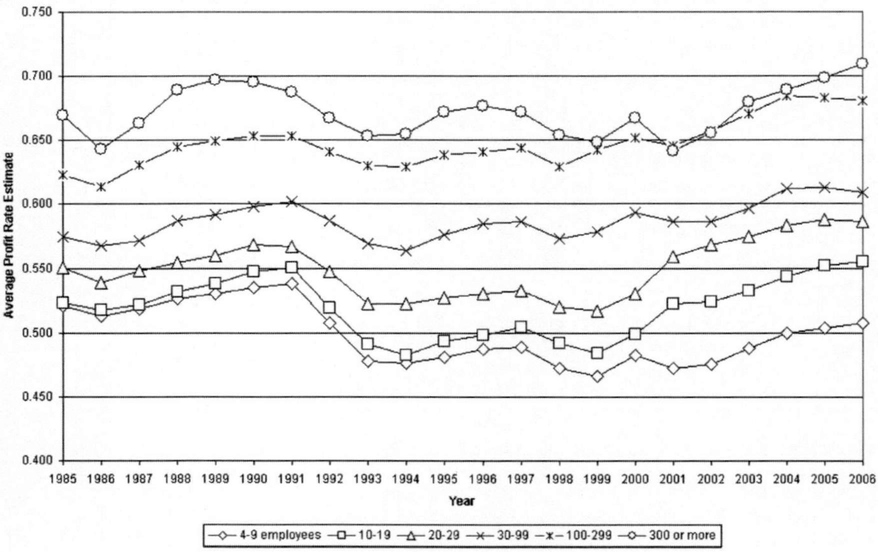

Fig. 7.2 Profitability trends by firm size category

$$T_w = \left\{ \sum_{i=1}^{n} \left(\frac{S_i}{S} \right) \sum_{p=1}^{k} \left(\left(\frac{S_{ip}}{S_i} \right) * \ln \left[\frac{S_{ip}/S_i}{V_{ip}/V_i} \right] \right) \right\} \tag{3}$$

where S is surplus value (Value added-wages), and V is value-added for place i and firm size category p.

For the second part of the analysis the local spatial clustering of prefectural contributions to inequality within each firm size category were examined using the Local Moran statistic built into ESRI's ArcGIS 9.3 software package. A local indicator of spatial autocorrelation (LISA), the Local Moran statistic provides a way to identify local clusters of both high and low contributions to inequality (Anselin 1995; Anselin et al. 2007). In addition, spatial outliers can also be identified (places with high contributions to inequality surrounded by places with low contributions, or conversely, places with low contributions surrounded by those with high contributions). The contributions to inequality for each firm size category were averaged across the 1985–1990, 1991–2001, and 2002–2006 time periods, and LISA cluster maps for each firm size category were produced using a first order spatial contiguity weights matrix.

In the third part of the analysis the LISA cluster maps were used to identify three core prefectures that demonstrated differences in their contributions to the overall levels of inequality. For these three core prefectures average profitability for their six firm size categories were examined relative to the national average levels of profitability. In addition, the relative ranking for each prefecture's six firm size categories were calculated.

7.4 Results and Interpretation

Figure 7.2 shows that estimates of profitability tended to be higher for the larger size firm categories, but generally all firm size categories experienced very similar trends during the study period. Not surprisingly the estimated levels of profitability trended upwards during the years of the bubble economy (1985 to 1990), which paralleled the trend in profitability for all firms in the late 1980s identified by Webber and Rigby (1996). However, profitability estimates dropped sharply in the early 1990s after the collapse of the bubble, with the strongest decline being particularly evident in the smaller size firm categories. From 1994 to 1997 all six firm size categories experienced a relatively mild upswing in profitability, possibly driven in part the efforts of the Japanese government to stimulate the economy through big-budget construction projects (Sorensen 2002). Possible effects of the Asian Financial Crisis can be seen in the decline of profitability estimates between 1997 and 1999. The strength of the economic expansion that started in 2002 is also evident in the chart. After a short decline from 2000 to 2001, all six of the firm size categories experienced rising profitability until the end of the study period in 2006. Most of the six firm size categories had higher profitability estimates in 2006 than for those of the peak years of the bubble economy, with the 4–9 full-time worker category being the only exception.

The relationship between the level of total inequality in average profitability levels between the firm size categories as measured by the Theil index and the raw trends in the average profitability varied during the period of the study. During the bubble years the peak in total inequality occurred in 1989 when profitability estimate for firms with over 300 regular workers moved sharply upward and the estimates for several other firm size categories were also rising (Fig. 7.3). In contrast, during the "lost decade of economic growth" of the 1990s total inequality peaked in 1994 and 1999, both occasions when the most of the average profitability estimates for the six firm size categories were at low levels. In the recovery of the early 2000s total inequality peaked in 2003 mainly due to high levels of inequality within firm size categories.

Further insight into these differences can be gained by focusing on the trends in the inequality between the firm size categories and the inequality within the categories (variation across prefectures). Figure 7.3 shows that total inequality during the bubble period was mainly driven by differences in profitability estimates between the firm size categories. While the inequalities arising from differences in profitability between firm size categories remained high during the 1990s, the inequalities within the firm size categories became more important, and peaks in total inequality occurred in those years in which it also spiked upwards. For the early 2000s the relationship shifted as inequality within firm size categories increased to the point where it was making a larger contribution to total inequality than inequality between the firm size categories.

It is clear from Figs. 7.4 and 7.5 that the largest size firm categories strongly influenced trends in inequality. The peaks in total inequality seen in 1989, 1994, and 1999 corresponded with peaks in the contribution of the largest firm size category

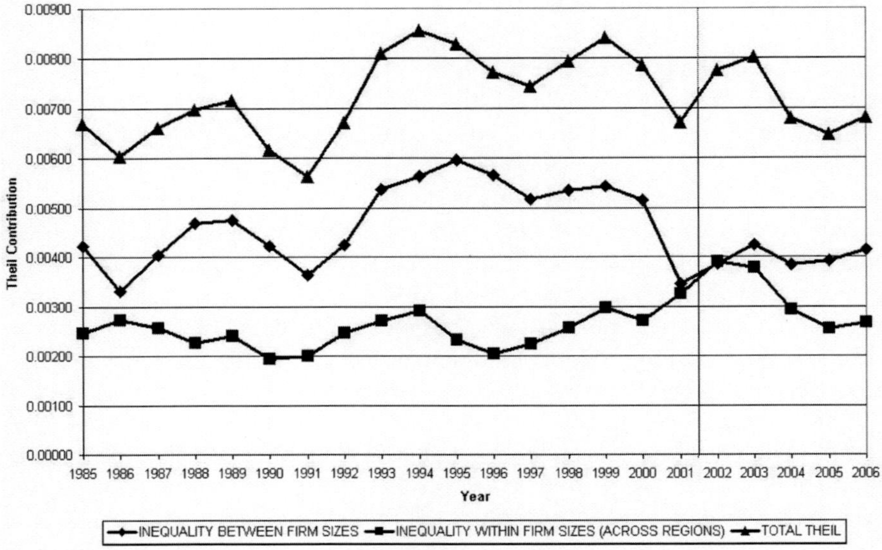

Fig. 7.3 Theil elements for firm size category profitability estimates

to regional inequality. In 2003 the pattern was slightly different with strong contributions towards regional inequality from both the 300 or more workers firm size category and the 100–299 workers firm size category. Also of note are the changes in the general trends of regional contributions to inequality starting in the late 1980s and early 1990s. The smallest firm size categories generally began to contribute to

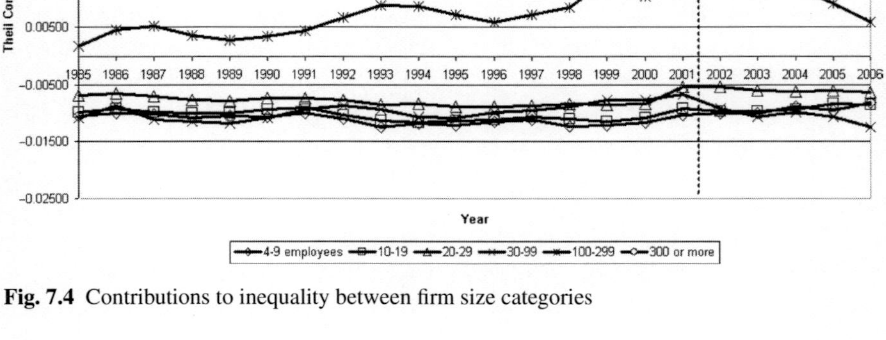

Fig. 7.4 Contributions to inequality between firm size categories

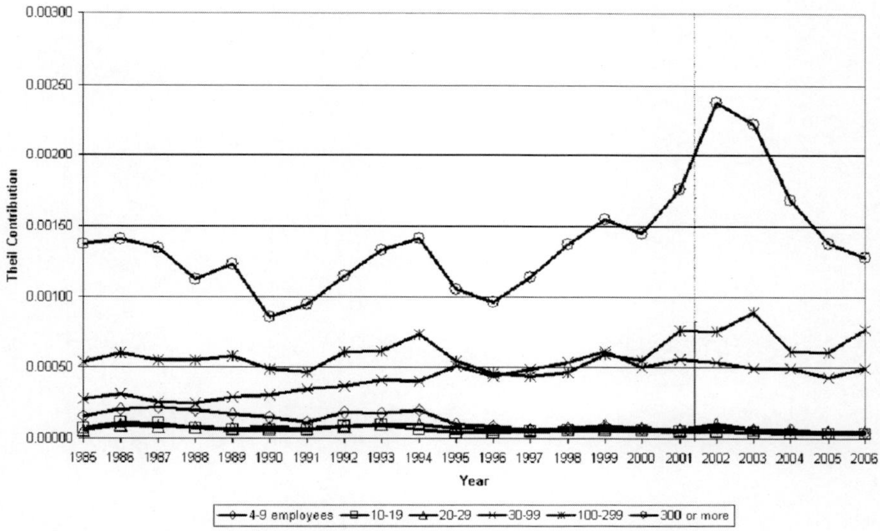

Fig. 7.5 Contributions to inequality within firm size categories

lower levels of regional inequality, while the larger firm size categories tended to trend upwards in the their contributions until about 2002–2003.

The LISA cluster maps for smaller firms shown in Figs. 7.6 and 7.7 suggest that during the bubble period statistically significant clusters of strong contributions to regional inequality are seen in the peripheral regions of the north and south, and significant clusters of the weakest contributions were located near Tokyo, Kanagawa and Saitama prefectures of the core manufacturing region. The spatial patterns for the two largest firm size categories are quite different with significant clusters of high contributions to inequality located in the core region and significant clusters of low values in the peripheral regions – particularly the northern periphery for the largest firm size category.

During the 1990s, contributions to regional inequality from the smaller firms demonstrate similar clustering patterns to those of the bubble years (Fig. 7.8). Once again significant clusters of high contributions are located in the northern and southern periphery, but are particularly strong in the north. One interesting change for both the 10–19 workers and 20–29 workers firm size categories is that Mie prefecture flips from being the center of a cluster of low contributions to regional inequality to become a spatial outlier with high contributions surrounded by prefectures with low contributions.

For the larger firm size categories there are more substantial differences between the two periods (Fig. 7.9). For the 30–99 workers firm size category the significant cluster of low contributions centering on Tokyo shrinks considerably, and the significant clusters of high contributions in the peripheral regions present during the bubble period are no longer significant during the 1990s. The 100–299 workers firm size category had significant clusters of high contributions centering on Tokyo

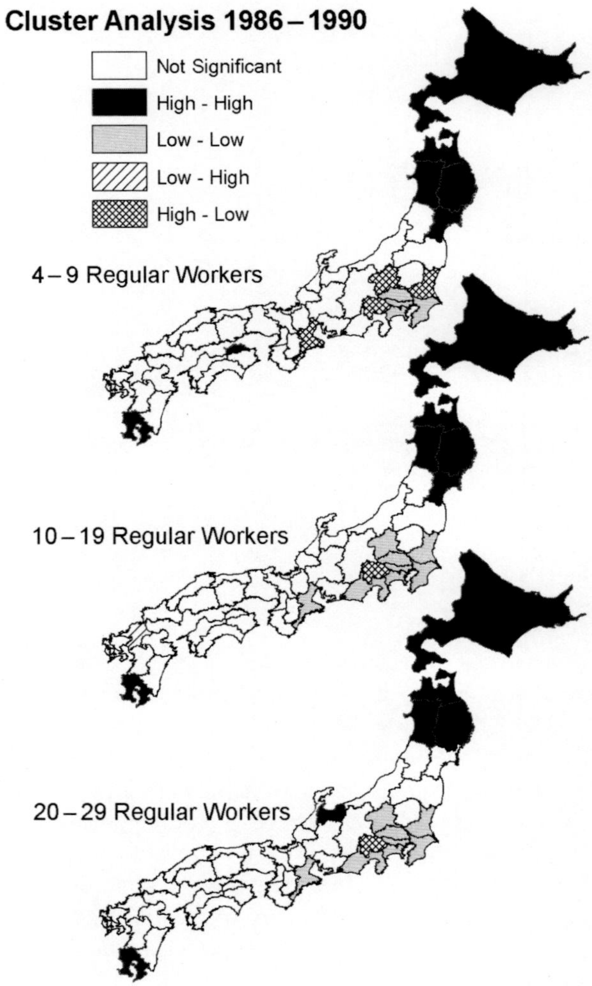

Fig. 7.6 LISA cluster maps of contributions to inequality 1986–1990 – smaller firms

and Osaka prefectures during the bubble years, but these clusters were considerably smaller during the 1990s, and many more spatial outliers are evident. The over 300 workers firm size category also has a smaller cluster of high contributions to inequality in the region surrounding Tokyo prefecture, and several clusters of high or low values present during the bubble period are no longer significant.

During the economic expansion period of the early 2000s most of the patterns for the smallest firm size categories saw only slight changes (Fig. 7.10). Only the smallest firm size category remains nearly identical to the previous pattern of the 1990s. The 10–19 workers category has a surprising change with Hokkai prefecture dropping out of the cluster of high contributions to become a low contribution spatial outlier. In addition, its cluster of significantly low contributions that centers on Tokyo became smaller. For the 20–29 workers category the significant cluster of low

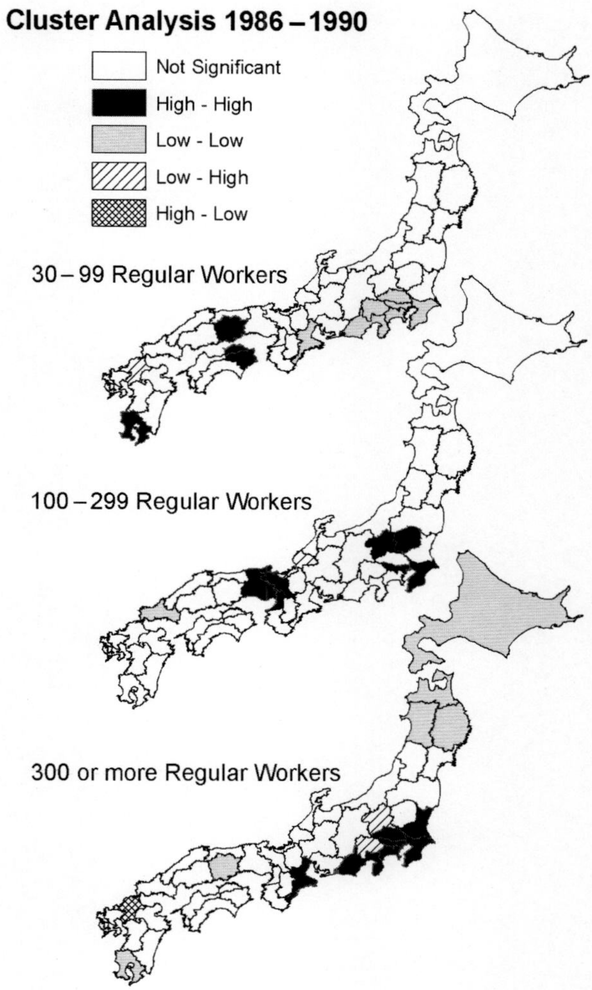

Fig. 7.7 LISA cluster maps of contributions to inequality, 1986–1990 – larger firms

contributions that centered on Tokyo changed slightly with Kanagawa prefecture no longer being significant, while Aichi prefecture became part of the low contribution cluster for the first time.

Two of the three larger firm size categories showed substantial changes in the spatial pattern of clustering during the early 2000s. The 30–99 workers category had little change, with a cluster of low contributions that included Tokyo and Chiba prefectures present during the 1990s losing significance during early 2000s. However for the 100–299 workers category a cluster of high contributions shift to include Tokyo prefecture, and several spatial outliers with low contributions are no longer significant. The pattern for the largest firm size category also has a substantial change with a large cluster of high contributions evident in the 1990s becoming insignificant in the early 2000s.

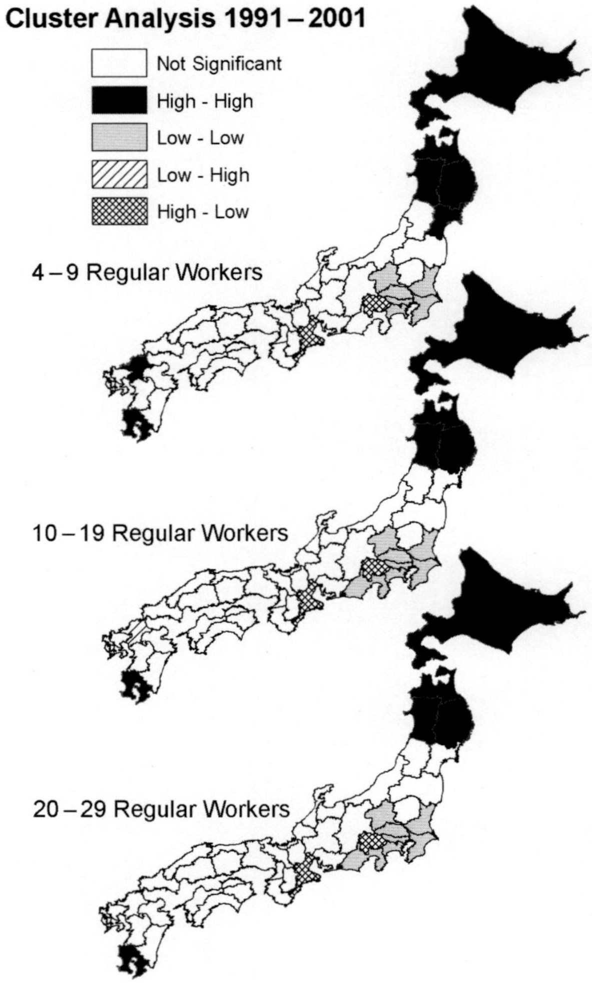

Fig. 7.8 LISA cluster maps of contributions to inequality, 1991–2001 – smaller firms

The LISA cluster maps suggest that much of the fluctuation in total inequality is tied to changes in the larger sized firms in the region surrounding Tokyo prefecture. The smaller firm size categories in both Tokyo and neighboring Kanagawa prefecture were often part of clusters of prefectures that made low contributions to overall levels of inequality in profitability. However, for the two largest firm size categories these same prefectures were often part of clusters that made high contributions to total inequality.

Many of Japan's small firm production networks have historically been concentrated in Tokyo prefecture, but Table 7.3 suggests that many of these firms have struggled to maintain viability. Although there is a sharp decline in the number of establishments for all firm sizes from 1985 to 2001, the largest declines were for the 4–9 workers category which dropped by nearly half. Even with the economic

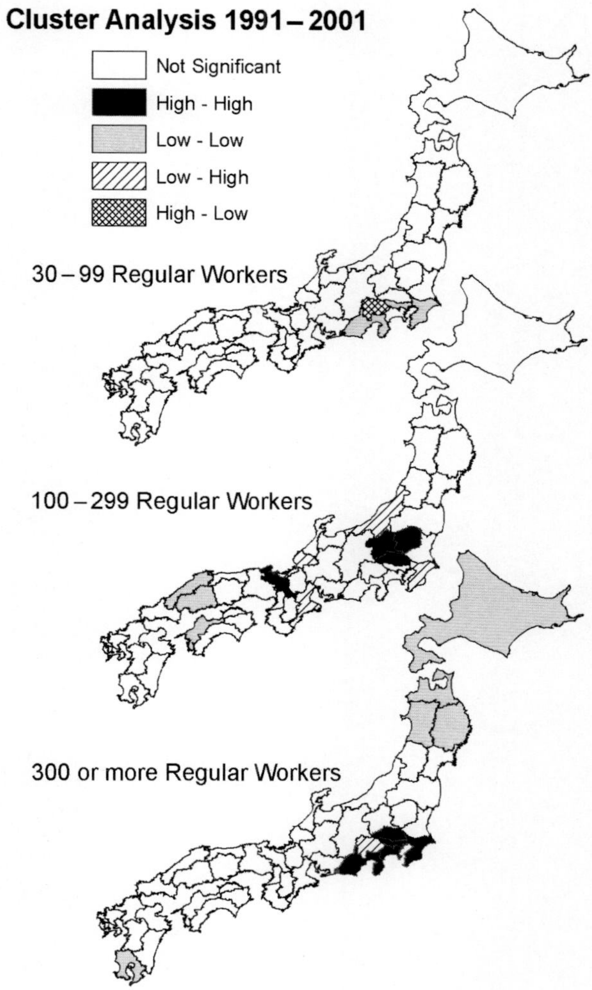

Fig. 7.9 LISA cluster maps of contributions to inequality, 1991–2001 – larger firms

expansion from 2002 to 2006 the number of establishments continued to decline, and most of them declining at a faster rate than during the 1990s.

Not surprisingly, the performance of Tokyo's small manufacturers was also quite poor when examined in terms of average profitability. Figure 7.11 shows that in Tokyo prefecture the largest sized firms tended to have higher than average profitability only during the peaks of growth periods (shaded markers on the chart indicate profitability estimates higher than the national average for that year), and in total were higher than the national average for only seven of the twenty-two years of the study. While most of Tokyo's firm size categories showed an upswing in profitability starting in the late 1990s, the relative rankings of Tokyo's firm size categories shown in Fig. 7.12 suggest that the upswing was relatively weak in comparison to the performance of manufacturers in other prefectures. Of particular note is the very

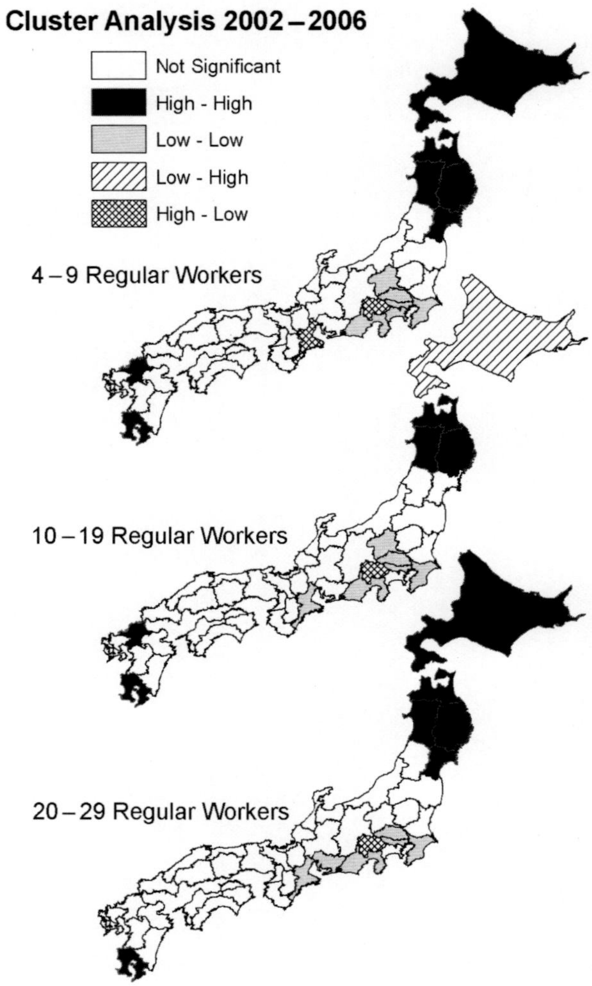

Fig. 7.10 LISA cluster maps of contributions to inequality, 2002–2006 – smaller firms

poor performance of Tokyo's smallest firm size category which consistently had the lowest profitability estimate of all 47 prefectures for the majority of the study period.

Although Aichi prefecture is also part of the core manufacturing region and has a large number of smaller sized firms, unlike Tokyo prefecture it is seldom identified as part of a significant cluster of high or low contributions to total inequality. Aichi prefecture also experienced sharp declines in the number of manufacturing establishments, although the losses were not as severe as for Tokyo prefecture (Table 7.4). Another contrast with Tokyo is that during the growth period from 2002 to 2006 the number of establishments in the largest firm size categories increased.

The average profitability of Aichi's firm size categories was relatively strong during the early years of the study period. Five of Aichi prefecture's six firm size categories had relatively strong profitability estimates during the bubble years, and

Table 7.3 Tokyo prefecture – change in establishments by firm size

Firm size (# regular workers)	1985 Est.	2006 Est.	Percent change 1985–2001	Percent change 2002–2006
Over 300	229	107	−33.2 (−2.1)	–4.5(−1.1)
100–299	679	283	−37.6 (−2.3)	−16.8(–4.2)
30–99	3,062	1,304	−42.9 (−2.7)	−12.2(–3.1)
20–29	3,760	1,752	−38.8 (−2.4)	−11.9(–3.0)
10–19	8,784	4,154	−35.7 (−2.2)	−16.9(–4.2)
4–9	33,383	11,438	−49.7 (−3.1)	−19.0(–4.8)
Total	49,897	19,038	−45.8 (−2.9)	−17.4(–4.4)

Note: Due to changes in Japan's industrial classification system, data after 2001 are not directly comparable to previous years. Average annual percent change is listed in brackets.
Source: METI Research and Statistics Division.

some of these also had higher than average profitability for several years in the 1990s (Fig. 7.13). In the early 2000s there is also a noticeable strengthening of profitability estimates for Aichi's three smallest firm size categories. This strengthening is also evident in the chart of relative rankings (Fig. 7.14). With the exception of the smallest firm size category, Aichi's categories were grouped closely together in the middle ranges during the bubble years. In the 1990s there was much greater variability in their rankings, and a slight downward trend, contrasting with Tokyo where many of the categories spiked upwards in their rankings during the same period. Interestingly, Aichi's smallest firm size category's ranking rises substantially during the last two years of the analysis so that all six categories are close in their relative rankings by the end of the study period.

Osaka prefecture, like Tokyo, has also historically had very strong production networks incorporating substantial numbers of smaller firms. Like Tokyo it too experienced declines in the number of establishments for all firm sizes during the study period (Table 7.5). In terms of profitability, however, smaller manufacturers in Osaka seem to be in a stronger position than those in Tokyo, particularly during the earlier years of the study period. The profitability trends for Osaka prefecture's

Table 7.4 Aichi prefecture – change in establishments by firm size

Firm size (# regular workers)	1985 Est.	2006 Est.	Percent change 1985–2001	Percent change 2002–2006
Over 300	339	364	–4.7 (–0.3)	14.5 (3.6)
100–299	798	844	–0.4 (–0.0)	8.8 (2.2)
30–99	2,791	2,372	–11.6 (–0.7)	–1.0 (–0.2)
20–29	2,929	2,378	–13.3 (–0.8)	0.5 (0.1)
10–19	6,042	4,979	–6.2 (–0.4)	–9.8 (–2.5)
4–9	22,480	10,800	–36.8 (–2.3)	–15.9 (–4.0)
Total	35,379	21,737	–26.5 (–1.7)	–10.2 (–2.6)

Note: Due to changes in Japan's industrial classification system, data after 2001 are not directly comparable to previous years.
Source: METI Research and Statistics Division.

Fig. 7.11 LISA cluster maps of contributions to inequality, 2002–2006 – larger firms

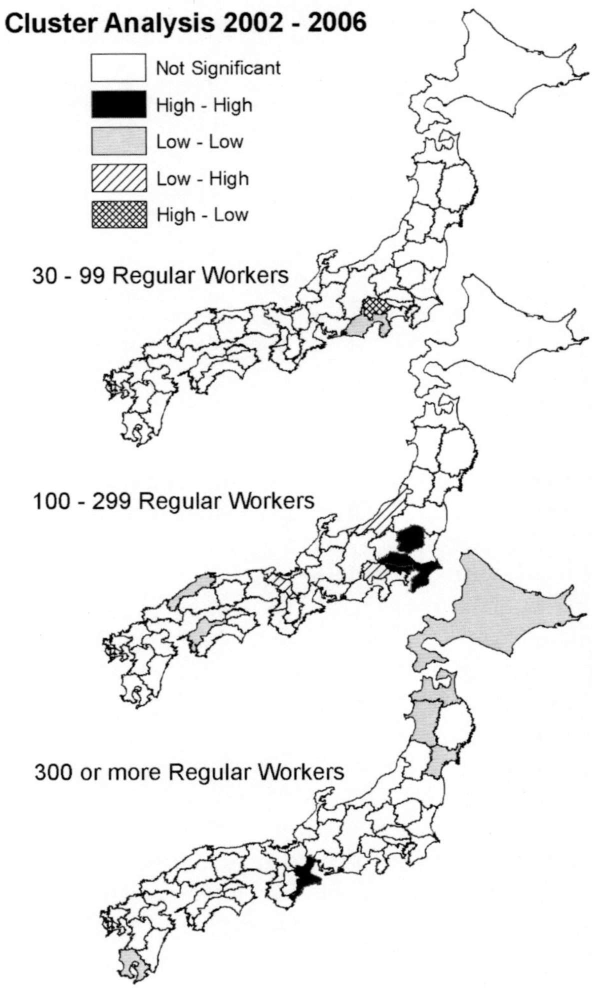

firm size categories are similar to those for Aichi prefecture. Profitability estimates for the bubble years are relatively high with all of the firm size categories except the largest having estimates higher than the national average (Fig. 7.15). Most of the categories experience a decline in their profitability estimates during the 1990s, but the 20–29 workers category remains relatively strong. Surprisingly, the chart of relative rankings for Osaka's firm sizes shows that the largest and smallest categories were the weakest and exhibited a downward trend until the early 2000s (Fig. 7.16). The other four categories were closely grouped at relatively high levels during the bubble years, but then gradually declined in their rankings until the end of the study period.

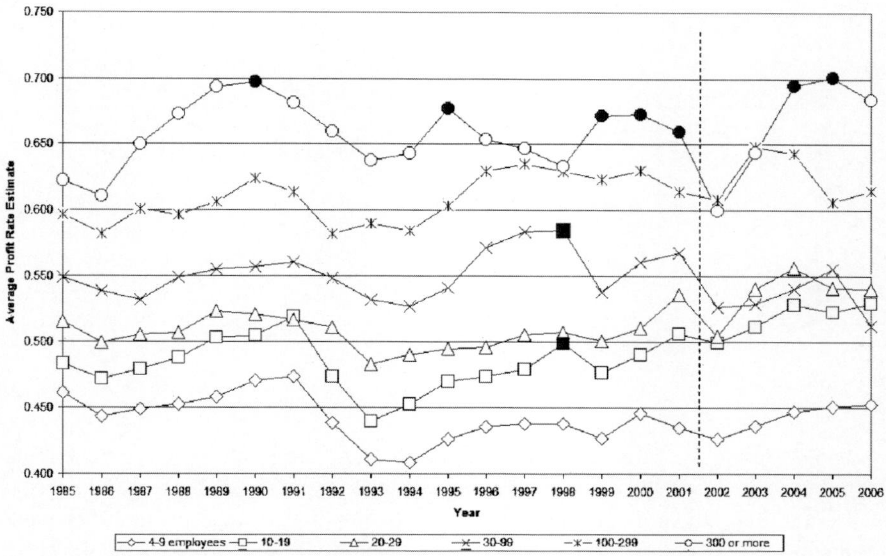

Fig. 7.12 Tokyo prefecture – profitability estimates by firm size category

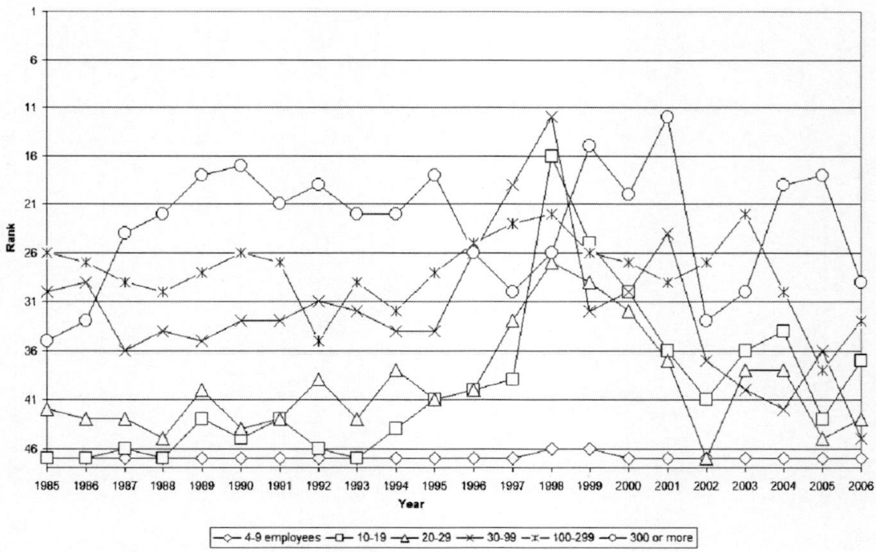

Fig. 7.13 Tokyo prefecture – firm size category profitability ranking

7.5 Conclusion

The analysis showed that there were substantial differences in the average rates of profitability according to firm size. However, it was also clear that the intense spatial and scalar restructuring that occurred during the study period was connected

Table 7.5 Osaka prefecture – change in establishments by firm size

Firm size (# regular workers)	1985 Est.	2006 Est.	Percent change 1985–2001	Percent change 2002–2006
Over 300	267	141	−31.8 (−2.0)	−15.6 (−3.9)
100 – 299	870	557	−28.2 (−1.8)	−1.8 (−0.4)
30 – 99	3, 311	2, 197	−25.2 (−1.6)	−6.8 (−1.7)
20 – 29	3, 900	2, 435	−29.5 (−1.8)	−2.8 (−0.7)
10 – 19	8, 921	5, 541	−25.2 (-1.6)	−10.4 (−2.6)
4 – 9	28, 545	12, 693	−40.7 (−2.5)	−16.1 (−4.0)
Total	45, 814	23, 564	−35.3 (−2.2)	−12.4 (−3.1)

Note: Due to changes in Japan's industrial classification system, data after 2001 are not directly comparable to previous years.
Source: METI Research and Statistics Division.

to changing long-term trends in average profitability. During the years of the bubble economy in the late 1980s the peak in inequality corresponded to high average levels of profitability, but in the 1990s peaks in inequality for average profitability were associated with economic downturns. As the economy began to recover in the early 2000s, the peak in inequality occurred during the middle of an upswing in average profitability levels.

The decomposition of total inequality into the contributions from inequality between categories and within categories suggests that much of the transformation in the long-term relationship was due to instability in average profitability for the larger firms. The contributions of the 100–299 workers category to between category inequality rose during the 1990s and when combined with the high levels of the over

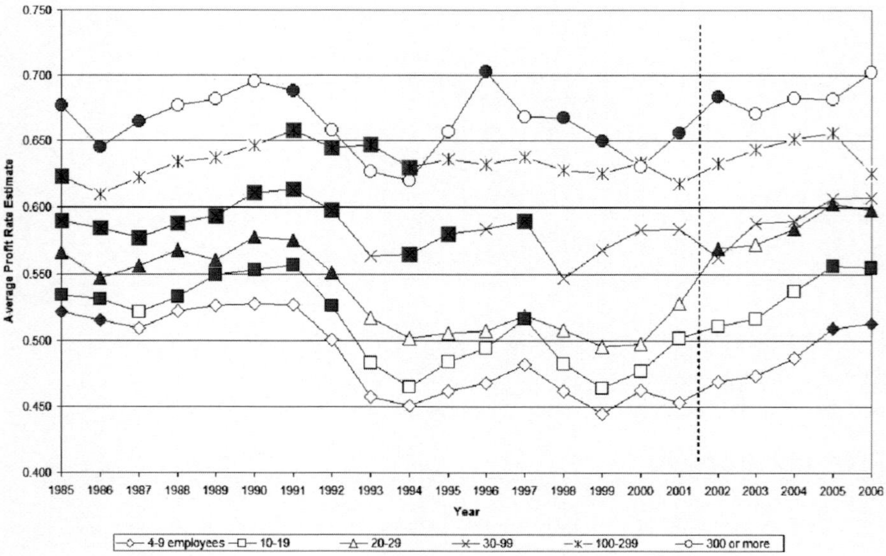

Fig. 7.14 Aichi prefecture – profitability estimates by firm size category

300 workers category made the main contribution to higher levels of total inequality during the economic downturns. In addition, the 2003 peak in total inequality was mainly due to a reversal of the downward trend in contributions to inequality by the largest firm size category.

The instability of average profit rates in the larger firm size categories is also evident in their contributions to regional inequalities. The contributions of the three largest firm size categories to regional inequalities increased for the majority of the study period, with particularly strong contributions from the over 300 workers category in the early 2000s. In contrast, the contributions of the three smallest firm size categories generally decreased after 1994.

The LISA cluster maps suggest that there are high levels of spatial variation occurring in the contribution to inequality within both the manufacturing core and periphery. For smaller firm sizes contributions were particularly strong in the northern prefectures of the periphery – Hokkaido, Aomori, Iwate, Miyagi, and Akita. The maps of the larger firm sizes often identified Tokyo and several surrounding prefectures in the northern section of the core as sites of strong contributions to regional inequalities. The examination of Tokyo, Aichi, and Osaka prefectures helped to illustrate the differentiation within the core region. Only in Aichi prefecture has the number of establishments increased for the larger firm size categories. Tokyo continued to lose establishments, and for most categories at a faster rate than during the late 1980s and 1990s. The largest size firms in Tokyo showed also showed larger fluctuations in relative average profitability compared to those in Aichi and Osaka.

One limitation of this research has been the lack of information on industrial structure. Although manufacturing sector data for the firm size categories are not

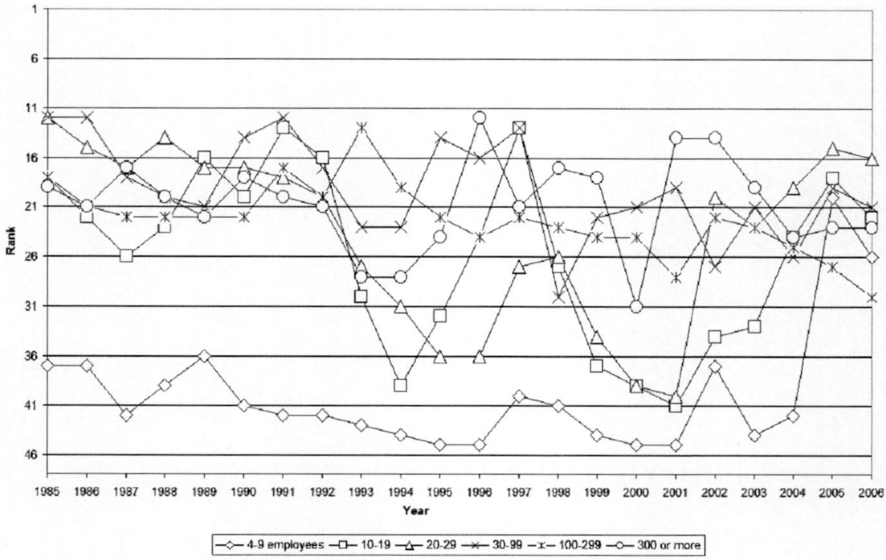

Fig. 7.15 Aichi prefecture – firm size category profitability ranking

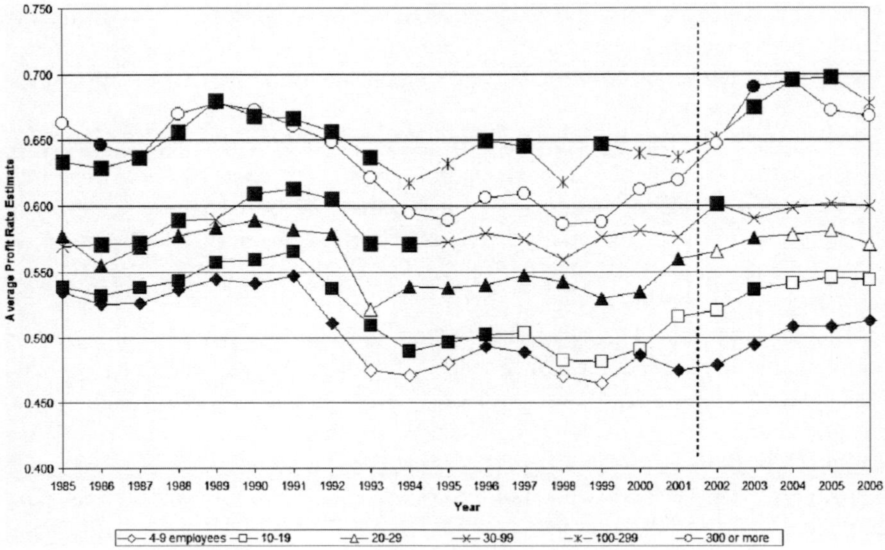

Fig. 7.16 Osaka prefecture – profitability estimates by firm size category

available, sector data for all firms could be incorporated in a more qualitative fashion, particularly in conjunction with the LISA cluster maps, to identify possible restructuring strategies in particular places. The research could also be strengthened by focusing on the metropolitan scale as well as the prefectural. Such improvements

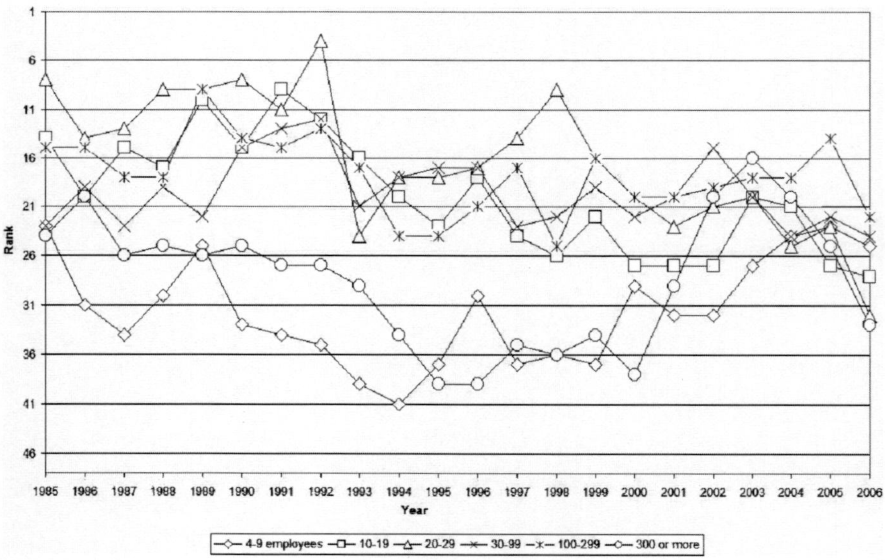

Fig. 7.17 Osaka prefecture – firm size category profitability ranking

could provide further insight into the ongoing spatial and scalar transformation of the Japanese industrial landscape.

References

Akita, T. (2003). Decomposing regional income inequality in China and Indonesia using two-stage nested Theil decomposition method. *Annals of Regional Science*, 37, 55–77.

Anselin, L. (1995). Local indicators of spatial autocorrelation – LISA. *Geographical Analysis*, 27, 93–115.

Anselin A., Sridharan, S., and. Gholston, S. (2007). Using exploratory spatial data analysis to leverage social indicator databases: the discovery of interesting patterns. *Social Indicators Research*, 82, 287–309.

Dehesh, A., and Pugh, C. (1999). The internationalization of post-1980 property cycles and the Japanese 'bubble' economy, 1986–96. *International Journal of Urban and Regional Research*, 23 (1), 147–164.

Fujita, M., and Hu, D. (2001). Regional disparity in China 1985–1994: the effects of globalization and economic liberalization. *Annals of Regional Science* 35, 3–37.

Galbraith, J. and Berner, M. (2001). Measuring inequality and industrial change. In J. Galbraith and M. Berner (Eds.), *Inequality and Industrial Change: A Global View* (pp.16–32). Cambridge: Cambridge University Press.

Galbraith, J., and Garcilazo, E. (2005). Pay inequality in Europe 1995–2000: convergence between countries and stability inside. *The European Journal of Comparative Economics*, 2(2), 139–175.

Galbraith,J., Krytynskaia, L., and Wang, Q. (2004). The experience of rising inequality in Russia and China during the transition. *The European Journal of Comparative Economics* 1(1), 87–106.

Glasmeier A., and Sugiura, N. (1991). Japan's manufacturing system: small business, subcontracting and regional complex formation. *International Journal of Urban and Regional Research*, 15(3), 395–414.

Harada, Y., and Onishi, S. (2003). Japan's great recession: what went wrong? Tokyo: Cabinet Office, Economic and Social Research Institute, ESRI Discussion Paper Series No. 77.

Harvey, D. (2001). *Spaces of Capital*. New York: Routledge.

Harvey, D. (2006). *Spaces of Global Capitalism: Towards a Theory of Uneven Geographical Development*. New York: Verso.

Hoshi, T., and Kashyap, A. (2004). Japan's Financial Crisis and Economic Stagnation, *Journal of Economic Perspectives*, 18(1), 3–26.

Kawai, H., and Urata, S. (2002). Entry of small and medium enterprises and economic dynamism in Japan. *Small Business Economics*, 18(1): 41–52.

Kimura, F. (2002). Subcontracting and the performance of small and medium firms in Japan. *Small Business Economics*, 18(1), 163–176.

Ministry of Economy Trade and Industry (METI) [previously known as Ministry of International Trade and Industry (MITI)] (various years). *Census of Manufacturers*. Tokyo: Research and Statistics Division, Minister's Secretariat.

Samuels, R., and Keller, W. (2003). *Crisis and Innovation in Asian Technology*. Cambridge: Cambridge University Press.

Schneider, F. (1991). Efficiency and profitability: an inverse relationship according to the size of Austrian Firms? *Small Business Economics*, 3, 287–296.

Smith, N. (1990). *Uneven Development: Nature, Capital and the Production of Space*. Second Edition. Oxford: Basil Blackwell.

Smith, N., and Dennis, W. (1987) The restructuring of geographic scale: coalescence and fragmentation of the northern core region. *Economic Geography*, 63(2), 160–182.

Sorensen, A. (2002). *The Making of Urban Japan: Cities and Planning from Edo to the Twenty-First Century*. New York: Routledge.

Statistical Research and Training Institute 2008, *Japan Statistical Yearbook*. Tokyo: Statistics Bureau, Ministry of Internal Affairs and Communications.

Webber M., and Rigby, D. (1996). *The Golden Age Illusion: Rethinking Postwar Capitalism*. London: Guilford Press.

Whittaker D. (1997) *Small Firms in the Japanese Economy*. Cambridge: Cambridge University Press.

Chapter 8
Situating Urban Environmental Risk: Using GIScience to Understand Risk in a Midwestern City

Trevor Fuller and Jay D. Gatrell

Abstract This chapter uses a GIScience approach toward assessing and correlating the relationship between urban environmental risk, environmental amenities and socio-economics. Using geographical weighted regression, the analysis demonstrates that socio-economic variables co-vary with observed environmental conditions.

Keywords Environmental justice · Geographically weighted regression (GWR) · GIScience · Socioeconomics · Midwest

Environmental justice is a field well-suited for the use of geo-technologies. Geographic information systems, remote sensing, and other technologies have been used in the recent past to not only assess the distribution of environmental risk, but also to decipher the politics of environmental risk within a broader conceptual and policy framework. The strength of GIScience lays in its ability to visualize environmental hazards. Visualization provides researchers and policy makers with new insights into the socio-spatial distribution of environmental risk across and among populations.

In this chapter, we build on the earlier work of Buzzelli and Jerrett, (2004) and Fuller and Gatrell (2007) to investigate the relationship that exists between socioeconomic status and environmental risk in Terre Haute, Vigo County, Indiana. Using GIS, remote sensing, census, and environmental data, the chapter presents a framework for assessing the often nuanced dynamics of environmental risk and socioeconomics throughout the urban and rural neighborhoods of Vigo County.

8.1 Study Area

The study area selected for this analysis is Vigo County, Indiana, located within the United States. Vigo County covers an area of 1,063 square kilometers, and holds as

T. Fuller (✉)
Department of Geography, University of Illinois at Urbana-Champaign, Urbana, Illinois, USA
e-mail: tfuller3@illinois.edu

J.D. Gatrell, R.R. Jensen (eds.), *Planning and Socioeconomic Applications*, Geotechnologies and the Environment 1, DOI 10.1007/978-1-4020-9642-6_8,

Fig. 8.1 Vigo County, Indiana (Fuller and Gatrell 2007)

its county seat and population center, the city of Terre Haute (Fig. 8.1, see Fuller and Gatrell 2007). Located along the east side of the Wabash River, Terre Haute, Indiana had a population of 69,614 in 2000, with an observed county wide median income of $33,184 and a median housing value of $72,500 (US Census 2002). Demographically, Terre Haute's population is predominantly white, with African-Americans constituting approximately 6.1%. The physical landscape of Terre Haute and Vigo County offer considerable variety with both dense and mixed urban, parks, suburban, and rural/agricultural regions present. This varied landscape offers a useful study area as such a region is often encountered throughout the Midwest.

8.1.1 Defining Environmental Justice

While research over the past few decades has revealed the dynamic nature of environmental risk, it seems the definition of environmental justice itself is rather dynamic as it has continued to change. Most often, environmental justice is loosely characterized by the notion that a particular subpopulation, often classified by race or income, is, or has been, enduring a disproportionate or unjust degree of environmental risk based primarily on their geography. This risk, referred to here as disamenities, refer to proxy variables for environmental quality. These can include facility locations, such as a facility that is listed as generating large quantities of hazardous waste, or sites which have a history of releases to the environment. As will be discussed later in this chapter, the selection of the most appropriate disamenity variables can, and is, often challenged for several reasons.

While environmental justice, as a field of both research and political activism, is often defined or labeled in many different forms (i.e., environmental racism, environmental equity, environmental injustice), the federal government formally defined vis-à-vis Executive Order 12898 in 1998. The Clinton order required the United States Environmental Protection Agency (US EPA) to consider environmental justice as part of their routine policy actions. To that end, the US EPA defines environmental justice as "the fair treatment and meaningful involvement of all people regardless of race, color, national origin, or income with respect to the development, implementation, and enforcement of environmental laws, regulations, and policies" (US EPA Office of Environmental Justice 2006). This definition functionally articulated two specific policy objectives: (1) process equity, referring to the ability of all people to voice their opinion regarding environmental issues related to proposed policy actions by the government; and (2) outcome equity, where a citizen is afforded the opportunity to seek recourse in response to environmental risk. In practice though, the implementation of environmental justice guidelines has been open to interpretation and the specific interpretation of fair treatment and meaningful involvement varies over time and across space.

In addition to the broad and varied definitions of environmental justice encountered in the literature, the exact relationship between socio-economics and risk and our understanding of this relationship continues to evolve and uncertainty exists. Indeed, investigations that examine the relationship between risk and socio-economic status often serve to thicken the haze surrounding the assertions that whether particular populations endure a disproportionate amount of environmental risk (Cutter et al. 2001; Buzzelli and Jerrett 2004). While some research has demonstrated a strong relationship between environmental risk and dwelling value, as well as lone-parent families, other research has indicated people of color were disproportionately exposed, especially working-class Latinos (Buzzelli et al. 2003; Pulido 2000). The dynamism of environmental risk is apparent as research continues to show the location-specific role of certain demographic/socioeconomic characteristics with regards to disproportionate environmental risk (Cutter et al. 2001). While some research has even revealed no direct relationship between minority populations and disproportionate environmental risk (Anderton et al. 1994), the majority of research continues to show some degree of disproportionate environmental risk faced by minorities and/or low income groups. The selection of variables to represent demographics/socioeconomic status presents the aforementioned variability in results. However, this does not account for all of the mixed or even contradictory results encountered in environmental justice research. In the following paragraphs we will examine the role of selecting an area of analysis in contributing toward variability in research findings.

The environmental justice literature provides numerous examples of the widespread and varied choice for a suitable area of analysis. Research continues to move across all areas and scales, from city-wide analysis (Mohai and Bryant 1992; Buzzelli et al. 2003; Buzzelli and Jerrett 2004; Pulido 2000) to the county, state, and national level (Margai 2001; Pastor 2001). As can be expected, the findings within environmental justice research have therefore not only been highly variable, but contradictory as well. One of the earliest studies investigating claims of environmental racism was the United Church of Christ case in which an examination of the spatial distribution of environmental risk was made at the level of zip code (United Church of Christ 1987). Ecological fallacy, in which the heterogeneity of a particular area of study is often missed due to the area being too large, certainly became a concern as study areas grew in extent (Anderton et al. 1994). Anderton et al. (1994) chose the census tract as the area of analysis to hopefully capture the heterogeneity present within the study area. In this chapter we use a smaller area of analysis, the US Census block group, which was the smallest geographical area for which critical Census data could be obtained.

8.2 Geo-techniques and Environmental Justice

The visualization and analytical capabilities of GIS render it an effective partner to environmental justice research. GIS provides a powerful source of information for policymakers and the public alike, with the ability to incorporate numerous socioeconomic and demographic conditions into any analysis of mere facility location or emissions data. Due to the increasing popularity and utility of GIS, many State and Federal Agencies have compiled, and continue to gather, spatial data, producing useful datasets for many different streams of geographic research. Government-gathered geographical data, especially environmental data, is heavily relied upon for research regarding environmental risk. In addition to GIS, remote sensing technologies have also become more readily available and have helped illustrate many interactions which had been previously missed. One example would be the relationship between quality of life and vegetation or 'greenness' (Gatrell and Bierly 2002). The combination of remote sensing and GIS produces a powerful spatial analytical capability that may lead to more effective policy outcomes (Jensen et al. 2004), or what Longley (2002) has referred to as better geographies.

While GIS and remote sensing provide an effective platform for visualizing environmental risk and socioeconomic status across a study area, there remains the challenge of statistically capturing the interstitial interactions among the variables. The statistical technique known as geographically weighted regression (GWR) offers a useful method for such statistical visualization. Whereas standard regression provides global statistics implying uniformity across space, GWR effectively calculates local statistics at regression points across a study area, which aids in visualization of phenomena (Fotheringham et al. 2002). Because of this, GWR was used here to assist in the capture of statistical changes in environmental risk across US Census Block Groups. GWR may also help to reveal previously unvisualized interactions that occur within and between space. Indeed, GWR has proven to be especially effective with respect to modeling the observed relationship between socioeconomic variables and observe environmental conditions (LaFary et al. 2008). For this reason, GIS, remote sensing, and GWR will be used as a collective suite of geotechnologies to effectively and perhaps elegantly assess questions of environmental risk to humans.

8.3 Data and Methods

The primary objective of this investigation was to assess the relative efficacy of both environmental quality data and the normalized difference vegetation index (NDVI) as proxy measures for socioeconomic conditions. On the following pages, we will discuss the methods used, including the rationale for the specific socioeconomic and environmental quality variables, the process of calculating NDVI, and the statistical techniques used.

8.3.1 Environmental Data Sets

The United States Environmental Protection Agency (US EPA) has required reporting of certain information under the guidance of environmental regulations for several decades. This information has provided extensive data sets for research relating to environmental quality. The first data set used for this investigation was the EPA's Toxics Release Inventory Program (TRI), which includes information regarding reported releases from regulated facilities throughout the United States. In particular, releases to air, soil, and surface water were used by first asking whether there has been a release, answered with a yes or no, and then adding the amounts released (air, soil, and water) to make one reported number or quantity. In this way, there was no differentiation between routes of release. Rather, the total amount of released contaminants from each facility or site is used. By not parceling out the release information by medium, we avoided an investigative slippery slope regarding the route of release, which leads to a consideration of the medium, meteorological, and hydrological factors.

Treatment, storage, and disposal facilities (TSDFs) database was the second environmental quality data source used. TSDFs are regulated under the EPA's Resource Conservation and Recovery Act (RCRA), which in part, was designed to monitor the flow of hazardous waste from generation through to the time of disposal, a process commonly referred to as "cradle to grave".

The third environmental quality data set included the locations of Superfund sites within Vigo County, Indiana. This data consists of sites that are currently on the US EPA's National Priorities List (NPL). Sites are placed on the NPL after regulatory officials investigate each site by following the Superfund cleanup process, beginning with notification to EPA of possible releases of hazardous substances. After each site is investigated it is either designated as needed no further remedial action or it is proposed for placement on the NPL.

TRI, TSDF, and Superfund data are location-based in their application to environmental justice research. The proximity of such facilities to particular communities or segments thereof is interpreted by many researchers as an indication of environmental risk, usually disproportionately distributed among the study area population. The TRI data was acquired from the US EPA via its online data download library. The information is provided in the form of ESRI shapefiles and associated files, which was imported into ESRI's ArcMap software for display and analysis. TSDF data and Superfund site data were acquired from the Indiana Department of Environmental Management via the Indiana Geological Survey's online GIS data download library.

8.3.2 NDVI

A NDVI map was created with the use of the remote sensing software ERDAS Imagine 8.7 (ERDAS) and a satellite-produced image of Vigo County. The satellite image was produced by Advanced Spaceborne Thermal Emission and Reflection Radiometer (ASTER) using the Terra satellite, yielding a spatial resolution of 15 m in the near-infrared and visible portions of the electromagnetic spectrum. Within ERDAS a NDVI was calculated by incorporating the near-infrared and red channels into the following formula:

$$\frac{\text{Near-infrared} - \text{red}}{\text{Near-infrared} + \text{red}}$$

NDVI is based upon the principle that the red (visible) portion of the electromagnetic spectrum is highly absorbed by chlorophyll present within plants or vegetation, while the near-infrared energy is reflected at high levels by a plant's mesophyll leaf structure (Tucker 1979). The calculated vegetation index then indicates the relative strength or reflectance of vegetation throughout the satellite image of the study area. A higher NDVI value indicates a more robust presence of vegetation. NDVI is unique in that it normalizes the various reflectance values by converting them to a value between -1 and 1 for each pixel in the image, with -1 representing no vegetation and 1 indicating robust vegetation. This investigation examines a NDVI of Vigo County, Indiana to determine its efficacy as a metric for socioeconomic status, and then compares the resulting capacity to that of the environmental quality data. Specifically, NDVI variables used in this analysis included the following:

- Standard deviation of NDVI values within a block group; Minimum NDVI value observed within a block group;
- Maximum NDVI value observed within a block group;
- Interaction of NDVI with population density; and
- Mean NDVI value

8.3.2.1 Socioeconomic/Demographic Characteristics

Specific socioeconomic and demographic variables selected for use in this investigation have been applied in much of the earlier research regarding environmental justice. Such variables are often used as indicators of socioeconomic status. The following socioeconomic and demographic variables were acquired from the US Census Bureau's online data library for the year 2000 and integrated into this analysis as indicators of socioeconomic status: (1) Median household income; (2) Percent African-American population; and (3) Percent female-headed households with at least one child.

8.3.3 Methods

This investigation uses three approaches: correlation, weighted least squares regression, and geographically weighted regression. Correlation—Pearson's R—was used to explore the relationships between variables and the significance of these variables. Using the Pearson's R results as a guide, weighted least squares regression models were tested using both ordinary least squares (or enter method) and step-wise approaches. The weighted least squares regression was performed using population density as the weighting variable.

Geographically weighted regression was used as standard regression statistical techniques often treat phenomena as occurring equally across a study area. As Fotheringham et al. (2002) discussed, spatial data often exhibit what has been termed spatial nonstationarity, or the nonuniform distribution of spatial information. The benefit of GWR in geographical research is that it accounts for unique characteristics of spatial data by calculating the necessary statistical measures at each point in the study area, which provides individual level or point-unique statistical information, allowing a researcher to identify disparities in the spatial distribution of various phenomena. GWR served this research well given that previous research has demonstrated the spatial nonstationarity of disproportionate environmental risk (Mennis and Jordan 2005). GWR and GIS are used to best capture the sinuous and nuanced position of environmental risk relative to socioeconomic status (Buzzelli and Jerrett 2004).

The model for this investigation included the following variables analyzed through OLS regression and GWR, as well as analysis using Pearson's Correlation between the Socioeconomic and environmental metrics.

The variables used were:

- Median household income
- African-American population
- Female headed households with at least one child
- US EPA toxics release inventory
- RCRA TSDFs
- Superfund sites

- NDVI minimum
- NDVI maximum
- NDVI mean
- NDVI standard deviation
- Population density

8.3.3.1 Interaction Terms

In addition to the variables listed above, an interaction term was created using the expansion method (Casetti 1972; Gatrell and Bierly 2002; Jensen et al. 2005). The expansion method developed by Emilio Casetti and challenged the simple assumption that observed statistical relationships remain constant across space. Specifically, Casetti (1972) attributed the inherently non-stationarity of spatial phenomena to the interaction of terms (e.g., varying relationships) across space. In this study, we utilize a single interaction between the mean NDVI value and observed population density for each block group within the study area to understand how the relationships might vary across space. Population density was used as it has been shown to be effective when modeling environmental parameters, such as NDVI, in an urban environment (see Heynen and Lindsey 2003; Talarchek 1990). The interaction term was created through the simple multiplication of two key terms or variables—in this case mean NDVI and population density.

8.3.4 *The Models*

The models presented were subjected to OLS, stepwise, and GWR. The study models are:

$$\begin{aligned} \mathrm{Y} &= \beta 0 + \beta \mathrm{TR}(u, v) + \beta \mathrm{TF}(u, v) + \beta \mathrm{S}(u, v) + \beta \mathrm{NSD}(u, v) \\ &\quad + \beta \mathrm{NMIN}(u, v) + \beta \mathrm{NMAX}(u, v) + \beta \mathrm{NMN}(u, v) + \beta \mathrm{I}(u, v) + \varepsilon(u, v) \end{aligned}$$

where:

Y is the dependent variable (socioeconomic status):
  Median household income,
  African-American population, and
  Female-headed households with at least one child
β0 is the constant;
  TR is the US EPA's toxics release data for Vigo County;
  TF is IDEM treatment, storage, and disposal facilities;
  S is the superfund facility data;
  NSD is the standard deviation of the NDVI values;
  NMIN represents the minimum NDVI value;
  NMAX is the maximum NDVI value;
  NMN is the mean NDVI value;

I refers to interaction terms using population density and NMN; and
ε refers to the statistical noise assumed to be present;

The general formula for GWR is:

$$y_i = \beta_0(u_i, v_i) + \Sigma_k \beta_k(u_i, v_i)x_{ik} + \varepsilon_i$$

where:

y_i is the dependent variable at location I;
β_0 is an independent variable;
(u_i, v_i) is the coordinate location for the ith point;
$\beta_k(u_i, v_i)$ is the function continuously measuring parameter values at each point I; and
ε_i is the noise associated with each point i

(Based on Fotheringham et al. 2002).

8.4 Findings

Spatial nonstationarity has plagued previous environmental justice research as it calls into question the effectiveness of statistical analyses of geographical relationships (Fotheringham et al. 2002). While the process of WLS does capture variability across space as driven by varying population densities, "global" WLS does so based on discrete points rather than across a continuous surface (Fotheringham et al. 2002). For this reason, GWR calculates local statistics, specifically local r-square values, to determine the model performance in "place" and across "space". In this chapter, we use local r-square values derived from GWR to visualize, or map, the spatial dynamics and model performance across the study area. WLS and GWR 3.0 yielded both global and local coefficients of determination. WLS regression was performed on the data, using population density as the weighting variable, in order to evaluate the Pearson's correlation values. We first examined the distribution of the relationships between socioeconomic conditions and environmental quality data using WLS regression. WLS indicated a very weak relationship between median household income and environmental disamenities. WLS was able to discern variability in that relationship across space within Vigo County, but the overall relationships were quite weak. Local r-square values generated within the GWR software were mapped to provide a visual reflection of the data (Figs. 8.2 and 8.3). GWR was used to determine whether there was spatial nonstationarity among the relationship(s) between socioeconomic conditions and environmental disamenities.

When examining median household income using WLS, all four of the environmental quality variables received correlation values of 0.07 or lower, with three of the four having negative values. The most closely correlated variables to median

Fig. 8.2 NDVI image of Vigo County, Indiana

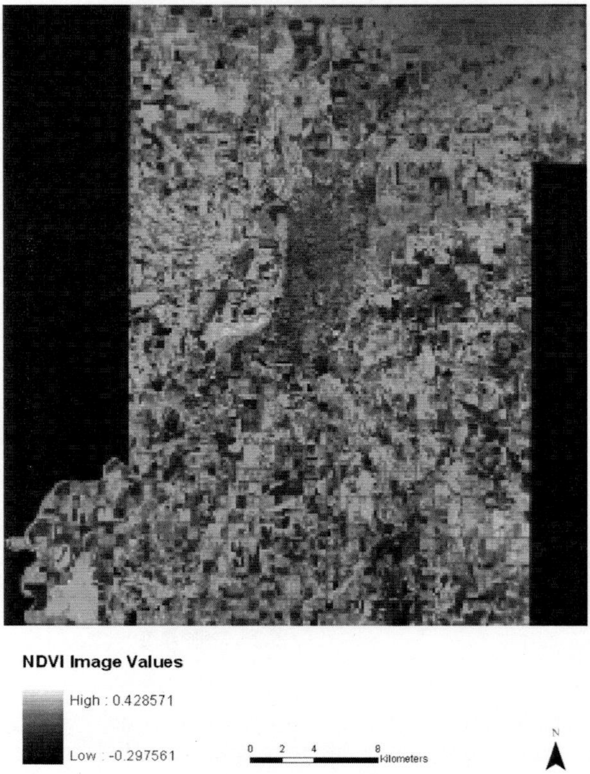

household income were the NDVI maximum value (0.439) and the NDVI mean value (0.557). With such a drastic disparity between the roles of environmental quality variables and NDVI variables, it is apparent that within Vigo County, the geographical distribution of traditional environmental quality indicators do not statistically account for the observed variation in and/or spatial arrangement of median household income.

As illustrated in Table 8.1, median household income was revealed as having an r-square value of 0.373. Among the independent variables used here, NDVI variables possessed a stronger relationship with median household income, with a significantly weaker showing for the environmental quality variables. These results challenge the historical treatment of certain forces behind environmental justice. The environmental quality variables (TRI, TSDF, and Superfund) do not appear to be significantly related to the distribution of such indicators of socioeconomic status as income and female head of household. These results may indicate that, rather than environmental risk being placed or cited within minority or low income areas, wealthier populations that may have been in these same areas have the ability (i.e., financial resources) to leave such areas and seek out areas with greater amounts of amenities, in this case, vegetation.

Fig. 8.3 Median household income local r-square values

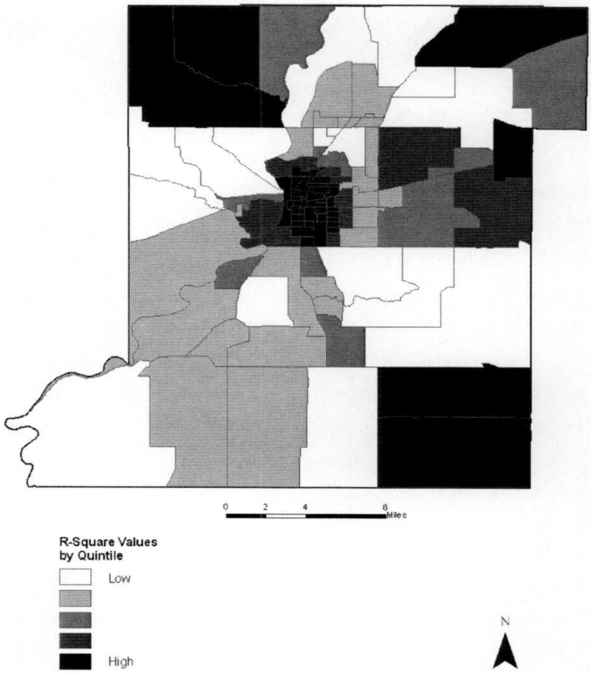

When the local r-square values for median household income, generated by GWR 3.0, were mapped (Fig. 8.3), there was a clear relationship within block groups of the urban core of Terre Haute between median household income and the independent variables used. This was not surprising as the urban center of Terre Haute contains block groups with some of the lowest income population. In addition, the model performed strongly in the northwest and southeast block groups, which are predominantly rural areas.

8.4.1 Race and Environmental Justice

Race has been a central concern in questions of disproportionate risk since the beginning of environmental justice both as a socio-political issue and a field of academic research. The majority of research conducted regarding environmental justice has included in its investigations, the role of race, and in particular the presence of actual or potential inequities endured by African-American populations. Here we consider the role of race in environmental (in)justice via inclusion of Vigo County's African-American population (Fig. 8.4).

By including consideration of the African-American population of Vigo County in this investigation, the model will conform to historical environmental justice research, which has often included the African-American population in order to investigation the role of race in environmental risk. As illustrated by the results, it appears the role of race here is minimal, with an R-square value of 0.135. Of particular significance, despite this low value, is the relationship between the African-American population and the interaction term (NDVI mean × population density), which was found to be significant at the 0.05 level.

8.4.2 Gender and Environmental Justice

Often missing from environmental justice investigations is the role of gender in regards to disproportionate risk. In this section, we consider the gender dynamics of environmental risk and model the observed relationship between the independent variables and the dependent variable female-headed households with at least one child (Fig. 8.5). The variable has been used as it is intended to capture not only gendered inequalities, but also other socio-economic aspects such as income which may also contribute toward disproportionate risk. Because of the presence of a university (Indiana State University) within the study area, there is an associated population of single females as heads of households. In an attempt to account for this place-specific limitation and avoid skewing the data, the analysis considers only female-headed households in which there is at least one child.

Table 8.1 Diagnostics (enter method)

Constant	Household income
	−6878.219
	(−0.494)
TR	−0.002
	(−0.456)
TF	−14122.071
	(−0.832)
S	2004.095
	(0.165)
NMin	30867.321
	(1.538)
NMax	75249.713
	(1.058)
NSDeviation	1128244.778
	(−0.911)
NMN	145973.149
	(2.468)†
NMN * Density Interaction	−22.483
	(−2.905)‡
R-Square	0.373
F-Statistic	7.148

* Indicates the variable is significant at the 0.10 level.
† Indicates the variable is significant at the 0.05 level.
‡ Indicates the variable is significant at the 0.005 level.

Table 8.2 Diagnostics (enter method)

Constant	African-American population
	57.587
	(0.488)
TR	−1.34E-005
	(−0.339)
TF	17.862
	(0.124)
S	4.414
	(0.043)
NMin	−59.396
	(−0.349)
NMax	380.300
	(0.631)
NSDeviation	−25.675
	(−0.022)
NMN	−862.300
	(−1.720)
NMN * Density Interaction	0.152
	(2.318)**
R-Square	0.135
F-Statistic	1.877

* Indicates the variable is significant at the 0.10 level.
† Indicates the variable is significant at the 0.05 level.

Table 8.3 Diagnostics (enter method)

Constant	FHH with at least one child
	36.886
	(1.289)
TR	−1.35E-006
	(−0.141)
TF	10.887
	(0.312)
S	8.986
	(0.360)
NMin	16.240
	(0.394)
NMax	−42.612
	(−0.292)
NSDeviation	360.858
	(1.248)
NMN	−115.602
	(−0.951)
NMN * Density Interaction	0.020
	(1.250)
R-Square	0.075
F-Statistic	0.972

* Indicates the variable is significant at the 0.10 level.
† Indicates the variable is significant at the 0.05 level.

8.5 Discussion

The environmental justice literature's focus on environmental disamenities seems to address a portion of the complex interactions that occur within and between social and natural systems in urban environments. Merely revealing the co-incidence of minority and/or low income populations and environmental disamenities, does not provide information sufficient enough upon which to base future policymaking decisions. A more complete story must be told, one which addresses questions regarding the longstanding implied relationships between negative externalities and class and race. The geography of such socioeconomic characteristics as class and race may be better understood via an assessment of access to environmental amenities, determined here by using NDVI as a proxy variable. Vigo County illustrates the broad range of socioeconomic characteristics over which disamenities occur, as there is a seemingly random spatial distribution of negative externalities throughout the urban and rural regions of the study area. With disamenities located in high and low income census block groups, as well as in predominantly non-African-American populated block groups, it's apparent that within Vigo County, class and race alone do not capture any degree of environmental injustice. The study suggests further

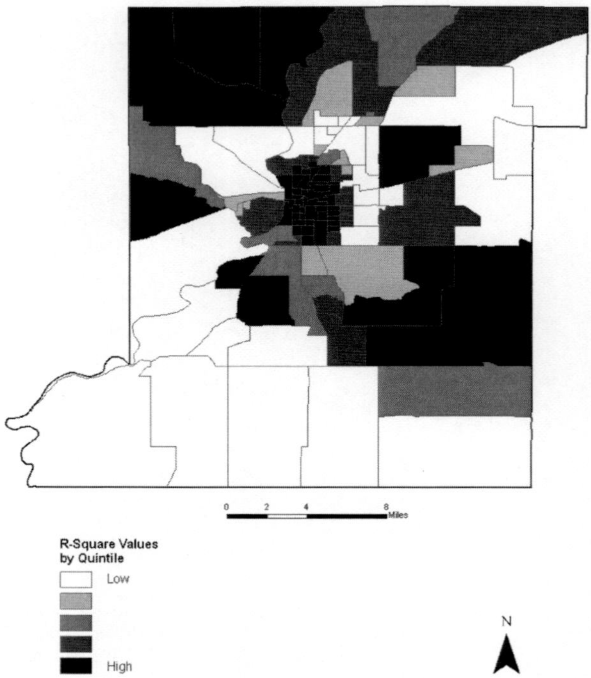

Fig. 8.4 African-American population local r-square values

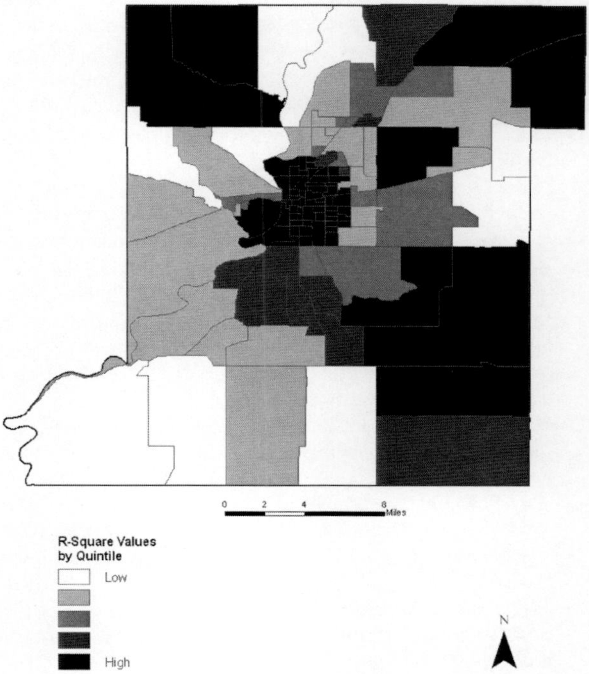

Fig. 8.5 Female head of household population r-square values

research is needed to understand the observed spatial disconnect between the urban geography of amenities and disamenities.

Acknowledgments The authors wish to acknowledge Eric LaFary for his technical assistance with GWR.

References

Anderton, Douglas L., A. Anderson, J. Oakes, and M. Fraser. 1994. Environmental Equity: The Demographics of Dumping. *Demography* 31: 229–248.

Buzzelli, Michael, M. Jerrett, R. Burnett, and N. Finklestein. 2003. Spatiotemporal perspectives on air pollution and environmental justice in Hamilton, Canada, 1985–1996. *Annals of the AAG* 93(3): 557–573.

Buzzelli, Michael and M. Jerrett. 2004. Racial gradients of ambient air pollution exposure in Hamilton, Canada. *Environment and Planning A* 36: 1855–1876.

Casetti, E. 1972. Generating models by the expansion method: applications to geographic research. *Geographical Analysis* 4:81–91.

Cutter, S., M. E. Hodgson, and K. Dow. 2001. Subsidized inequities: The spatial patterning of environmental risks and federally assisted housing. *Urban Geography* 22 (1): 29–53.

Fotheringham, A. Stewart, C. Brunsdon, and M. Charlton. 2002. *Geographically Weighted Regression: The Analysis of Spatially Varying Relationships*. John Wiley & Sons, Ltd. West Sussex.

Fuller, T., J. Gatrell, and E. LaFary. 2007. The spatial imperatives of environmental justice: Or, using geo-technologies to unlock the urban geographies of disamenities and amenities. In Jensen, Gatrell, & McLean, *Geo-Spatial Technologies in Urban Environments: Policy, Practice, & Pixels, 2nd Edition*. Heidelberg, Germany: Springer-Verlag. pp. 215–230.

Gatrell, J.D. and G.D. Bierly. 2002. Weather and voter turnout: Kentucky primary and general elections, 1990–2000. *Southeastern Geographer* Vol. XXXXII (1): 1–21.

Gatrell, J. and R. Jensen. 2008. Socio-spatial applications of remote sensing in urban environments. *Geography Compass* 2(3): 728–743.

Gatrell, J. D. and R. R. Jensen. 2002. Growth through greening: developing and assessing alternative economic development programs. *Applied Geography* 22: 331–350.

Heynen, N., and G. Lindsey. 2003. Correlates of urban forest canopy: implications for local public works. *Public Works Management and Policy* 8: 33–47.

Jensen, R., J. Gatrell, J. Boulton, and B. Harper. 2004. Using Remote Sensing and Geographic Information Systems to Study Urban Quality of Life and Urban Forest Amenities. *Ecology and Society* 9(5): 5. http://www.ecologyandsociety.org/vol9/iss5/art5/

Jensen, R., J. Gatrell, and D. McLean (eds). 2005. *Geo-Spatial Technologies in Urban Environments*. Heidelberg, Germany: Springer-Verlag.

LaFary, E., J. Gatrell and R. Jensen. 2008. People, pixels, & weights in Vanderburgh County, Indiana: Toward a new urban geography of human environment interactions. *Geocarto International* 23: 53–66.

Longley, P. A. 2002. Geographical information systems: will developments in urban remote sensing and GIS lead to 'better' urban geography? *Progress in Human Geography* 26:231-239.

Margai, Florence L. 2001. Health Risks and environmental inequity: a geographical analysis of accidental releases of hazardous materials. *The Professional Geographer* 53(3): 422–34.

Mennis, J. and Lisa Jordan. 2005. The Distribution of Environmental Equity: Exploring Spatial Nonstationarity in Multivariate Models of Air Toxic Releases. *Annals of the Association of American Geographers*, 95(2): 249–268.

Mohai, Paul, and Bunyan Bryant (eds.) (1992). *Race and the Incidence of Environmental Hazards: A Time for Discourse*. Boulder, Colo.; Westview.

Pastor, M., J. Sadd, and J. Hipp. 2001. Which came first? Toxic facilities, minority move-in, and environmental justice. *Journal of Urban Affairs* 23: 1–21.

Pulido, L. 2000. Rethinking environmental racism: white privilege and urban development in southern California. *Annals of the Association of American Geographers* 90: 12–40.

Talarchek, G. 1990. The urban forests of New Orleans: an exploratory analysis of relationships. Urban Geography 11: 65–86.

Tucker, C.J. 1979. Red and photographic infrared linear combinations for monitoring vegetation. *Remote Sensing of the Environment* 8: 127–150.

United Church of Christ Commission for Racial Justice. 1987. *Toxic Wastes and Race in the United States: A National Report on the Racial and Socioeconomic Characteristics of Communities with Hazardous Waste Sites*. New York: Public Data Access.

US Environmental Protection Agency. 2006. Toxics Release Inventory Program. www.epa.gov/tri/

US Environmental Protection Agency. 2006. Office of Environmental Justice. www.epa.gov/compliance/basics/ej.html.

Chapter 9
GIS and Challenges to Planning and Development Applications in Peripheral Regions

Rajiv Thakur and Madhuri Sharma

Abstract Planning and development activities in peripheral or non-urban spaces have undergone profound changes under the influence of geographic information systems (GIS) and spatial data analysis in the last two decades. This chapter investigates the challenges (methodological and practical) in the implementation of GIS to economic and planning applications in the peripheral regions (the developing and the developed world) in the context of society's transition from paper maps to digital geographic information. This research is significant since it is premised that implementation of GIS itself is 'contextual' and involves an appreciation of the human conditions(s) experienced in different places. The findings of this chapter's investigations as a result of review of literature is that much of the challenges lie not as much in how to introduce the technology to recalcitrant institutions in the peripheral regions of the world but in managing the availability of a mature technology and empowering the people while embedding the technology through participatory GIS in the context of different places. In other words the challenges are not simply technical but they are also political in the broadest sense of the word.

Keywords GIS · Planning · Peripheral Regions · Developing Nations

9.1 Introduction

In the last two decades advances in information infrastructure broadly, dramatic improvements in computer hardware and software and spectacular growth in tools for high quality cartography, three-dimensional visualization, analyzing networks or computing statistical measures of spatial patterns, along with the increased availability of internet infrastructure and the world wide web has all resulted in the proliferation of the use of GIS as a spatial decision support tool in economic planning and

---

R. Thakur (✉)
Department of Earth Sciences, University of South Alabama, Mobile, AL, USA
e-mail: rthakur@usouthal.edu

J.D. Gatrell, R.R. Jensen (eds.), *Planning and Socioeconomic Applications*,
Geotechnologies and the Environment 1, DOI 10.1007/978-1-4020-9642-6_9,

development even in the peripheral regions of the developed and developing world (Anselin 2000; Curry 1998; Goodchild and Haining 2004; Sui 2004). The 1970s and 1980s witnessed fundamental economic restructuring in western economies involving deindustrialization and lay-offs, restructuring in methods of production and the emergence of new forms of international economic organizations which suggested important changes in the way spatial planning was dealt with and theorized. GIS was used to facilitate a variety of management and planning decisions in both the public and private sectors of the peripheral regions (Gatrell 1999; Hall 1999). There is not much quantitative data to assess the penetration of GIS and the emerging challenges in the peripheral regions of the world. However, a review of literature shows that there is fundamental difference in the nature of applications of GIS in the socioeconomic domain in the peripheral regions of the developed and the developing world (Curry 1995; Hall et al. 1997, Problems and Prospects for GIS-based Decision Support Applications in Developing Countries. Unpublished paper, University of Waterloo; Hall 1999; Kyem 2001). For instance, peripheral regions in the United States benefit from the significant technological innovations and the development of topologically structured geographical referencing in the Dual Independent Map Encoding or the DIME system (Drummond 1995). Besides, the growth of application of GIS in the socioeconomic domain of the peripheral regions of the developed world has been facilitated by parallel advancements in computer operating systems, computer graphics, database management, computer-human interaction, graphical user interface design, and object-oriented programming methodologies. Some peripheral regions in developing countries such as India, China, Malaysia and South Africa among others have evolved in recent years as nascent spatial information technology infrastructure that is capable of supporting a high level of spatial information technology use (Yapa 1991). However, these are exceptions than the rule. Thus, the context and concerns of GIS applications in the socioeconomic domains of peripheral regions of the developed and developing countries varies.

This chapter is intended as a contribution to this purpose. The purpose is not to provide the reader with a complete and detailed overview of challenges but rather to focus on the differing contexts. Therefore, it presents opportunities available to GIS applications in economic and planning sectors of the peripheral regions and the challenges both methodological and practical emerging therein.

The chapter has the following structure. First, it separately assesses existing GIS capability to understand how peripheral regions capitalize on preexisting contexts and opportunities as is presented by both the expansion of the technology itself and the growing demands of economic restructuring. This is done merely through an examination of the literature. The idea is not to go into the potential debates but highlight the preeminence of 'context'. Second, we outline the challenges as they unfold from the examination of differing contexts. In fact, every place provides its own specifics making the implementation of GIS a heterogeneous activity across peripheral regions of the world. The conclusions argue for a critical engagement in the applications of GIS in the peripheral regions.

9.2 Unpacking the Context

A key element in applications of GIS in the context of developed and developing country's peripheries is how deeply embedded the technology tends to be in its economic, political and cultural milieu. GIS as a technology has interacted with evolving spatial urban planning issues of quality of life, competing claims on space and movement, management of degradation and equity and the need for urban development and growth. GIS is in many ways the result of adapting generic technologies to the particular needs of spatial data.

9.2.1 Developed Countries

Since the late 1970s and early 1980s economic restructuring and its impact on landscapes across the developed world has transformed the manner in which GIS has been used. In the US and Canada, applications of GIS has been influenced by changes in urban spatial planning and practice. Intrinsic to the growth of GIS in the US and Canada was the emergence of US Census Bureau led Topologically Integrated Geographic Encoding and Referencing (TIGER) System and the Government of Canada led Canada Geographic Information System respectively. These government sponsored projects consolidated the much desired need for integrating information and data in the design of software and map-making. The movement for urban re-development of 70s and 80s and the resultant re-articulation of urban space has also influenced GIS applications to planning practice (Nijkamp and Scholten 1993). In European countries the realization of severe demands on public space and the cry for sustainable development has brought in the involvement of public participation and engagement in spatial planning and decision making (Steinmann et al. 2004). The Australian experience of GIS development has been more in response to the need to handle geographic data pertaining to biophysical resources, social services in urban areas, planning and economic development, transportation, utilities and infrastructure, and especially land ownership, land titles, registration and land taxation (Garner and Zhou 1991). In a sense GIS did to planning practice in 1970s and 1980s what the 'quantitative revolution' in the 1960s did to spatial analysis in geography (Goodchild and Haining 2004). More specifically, GIS allowed the examination of planning and development applications to be understood from a fresh perspective particularly when notions of 'scale' and 'region' itself were being re-examined under the influence of economic restructuring (Massey 1994).

While the 1970s saw GIS infrastructure grow for its planning and development applications, the 1990s witnessed a second revolution in the application of GIS to planning practice with dramatic improvements in computer software and hardware and the internet explosion (Brail and Klosterman 2001; Goodchild and Haining 2004). Some feel that GIS as technology has come to determine contemporary planning practice (Nijkamp and Scholten 1993). However, in the developed countries this capacity of GIS to appropriate and determine planning and development application has come to be challenged by the almost paralleled growth of a critical discourse on GIS in the last decade under the rubric of ontology, representation, scale and more recently with the growing literature on collaborative decision making and public participation (Jankowski and Nyerges 2001; Kwan 2002; Lloyd et al. 2002; Schuurman 2006; Steinmann et al. 2004; Sui 2004). Thus GIS as a technology has assisted communities to come forward and participate in decision making by engaging in the deconstruction of existing spatial planning process itself.

9.2.2 Developing Countries

While governments in developed countries and their peripheries have pushed forward GIS and spatial data analysis into the forefront, drastically changing not just how information and communication is managed relative to citizens, but in ways also facilitating a critical engagement through the encouragement of public participation. Developing countries on the other hand, while having accepted in theory and practice the need to propel GIS, have in some places failed to appreciate and understand and hence in some places failed to manage and integrate critical organizational resources in such a way that GIS as a technology can be put in place to effectively assist in spatial decision support (Hall et al. 1997). A review of literature shows that the same processes of economic restructuring which entrenched GIS into the spatial decision support system in developed countries and their peripheries throughout 1980s and 1990s, prepared different parts of the developing countries and their peripheries differently in absorbing, assimilating and installing the tools and infrastructure required to adopt spatial decision support technology based upon there level of integration into the world markets with their local economies (Maguire et al. 1991). Thus, countries such as China and India on the one hand and Indonesia, Brazil, Malaysia, Thailand etc. on the other with such spectacular and rapid rates of urbanization are far ahead in their GIS applications and related infrastructural building needs than when compared to many other developing countries such as Vietnam, Egypt, Botswana, Ghana, Chile or for that matter Costa Rica, Mali, Cameroon, Qatar, Oman, Lebanon etc. (Hall et al. 1997). The installation of spatial decision support infrastructure including human capital and technical components requires funding which often most developing countries cannot afford and hence tend to lag behind in their preparedness for the application of GIS in planning and development domains. However, some developing countries such as India, Ghana, Pakistan, Bangladesh, Malaysia, Indonesia and Brazil have managed to build appropriate GIS information system infrastructure and tools with initial support through external funding and development aid projects (Hall 1999). Thus, there are disparities in the availability of GIS applications infrastructure in areas such as Sub-Saharan Africa, the Middle East and similar places in the South and Central America too.

Whatever be the context in the peripheral areas of developed and developing countries of the world, the significance of GIS is compelling and emerging economies such as China, India, Brazil, Thailand and Malaysia among others through their integration into the world markets have set the tone for collaborations and diffusion of the technology for greater spatial planning and decision support applications.

9.3 Challenges

Given the context, the challenges in the implementation of GIS applications to planning and development domain have evolved along different paths in peripheral regions of the developed and developing countries. Even the literature illustrates that GIS applications in the peripheral areas have lagged behind the metropolitan areas or urban spaces (Martin 2003; Schuurman 2004).

9.3.1 Methodological Challenges

While the rest of this section consolidates and addresses broad challenges under two heads namely methodological and practical, a number of these challenges overlap as methodological, database or scalar issue, all at the same time.

9.3.1.1 Data

As the use of GIS in the public sector has increased over the last couple of decades, and as new applications of the software vary in sophistication, whether one is an urban planner in a remote peripheral area of the United States or India or a user from the crime prevention, health care, environmental policy and traffic control or the higher education department the primary challenge is data accessibility. As peripheral areas begin to incorporate digital geographic data into everyday usage, accurate measurement and reporting of space has become important to patrons, users and institutions which are data repositories. As was experienced in Ghana, where steps taken to adopt GIS in the administration of land ran through the challenge of paucity of reliable data sets and lack of standards (Karikari et al. 2005). Ghana's problems and challenges with opportunities for use of GIS applications are not unique. A similar problem of poor spatial data infrastructure was also experienced in Botswana where the country's leadership made an attempt to improve information for land administration and management (Nkambwe 2001). These problems can well be replicated under similar situations in other African countries that share similar history, culture politics, economies, needs and resources (Hastings and Clark 1991). According to Schuurman (2004) "the devil is the data". Unfortunately, institutions that incorporate digital geographic data into everyday usage, quite often run into 'mystery data'. In particular, peripheral areas of developing countries have poor GIS data documentation or quite often the same is absent for it might have been developed for a specific project with no thought of subsequent uses or users.

With many federal, state and local agencies in the US now developing GIS datasets for spatial features such as roads, highways, rivers, lakes and such other landscape features, these spatial datasets also contain additional attributes in the form of demographic or socio-economic data of the region, becoming extremely useful in strategic decisions dealing with business, government and social issues of the peripheral regions (Drummond 1995). While, in countries such as the US and its peripheral areas data quality may not be a challenge, since there are rigorous procedures for compiling, creating, printing, and distributing spatial data and maps, in some peripheral regions of developing countries the accuracy of published data do not undergo a rigorous validation process.

9.3.1.2 Methods

A typical GIS analyst faces various challenges whether in developed or developing countries while trying to incorporate non-spatial dataset to spatial dataset in order to present resulting dataset on a geographical map. In other words does the practitioners employ appropriate techniques? Intrinsic to such methodological issues are assumptions about data representation and types of questions asked by the analyst (Csillag and Boots 2005). In so much the challenges of living with GIS are not simply technical, but they are also political in the broadest sense of that word. Hoeschele (2000) provides a great example from Kerala in India where state agencies abuse the power of GIS over those people who have little or no access to the technology. In a remote hamlet of Attappadi in the Western Ghats of Kerala, land cover data has been substituted for land-use data. As Hoeschele reports from intensive field studies that satellite vegetation cover data greatly overstates the amount of wastelands in the region. This is a classic example of a case where the productive role of the local indigenous population in managing land is grossly underrated. However, this kind of abuse of 'geographic information engineering' as Hoeschele calls it could be rectified by technically adding layers of data on socially specific land uses and institutionally by a more responsible public participatory method and process of doing GIS.

9.3.1.3 Scale

Since the move from representing the real world on paper maps to the digital medium is essentially about scale among other things, the issue of scale presents fundamental problems of cognition, measurement, representation and presentation of geographic information and how they translate into the digital environment (Lam and Quattrochi 1992; Quattrochi and Goodchild 1997). The issue of scale has also emerged problematic in GIS application in peripheral areas primarily because of the reconceptualization of geographic scale itself in urban geography (Agnew 1993; Cox 1998; Herod 1991; Jonas 1994; Massey 1994; Smith 1992). Urban and social geographers have been concerned that lack of access to GIS technology, databases and decision-making has often marginalized vulnerable groups in the peripheries of developing countries. The way GIS is constituted and practiced tends to isolate the lay-persons from participating in ongoing policy and planning debates (Curry 1995; Hoeschele 2000).

9.3.1.4 Representation and Visualization

The Attappadi case discussed above exposes the challenges that GIS applications to planning and development brings at it represents real-world geographic phenomena, human understanding of those phenomena and their representation within digital databases (Hoeschele 2000). Already many geographers, cartographers and GIS analyst have come to accept that cartography is at crossroads with the expansion of digitized map making. Essentially the business of map making is about representation which in turn is also about communication of phenomena in space. As such in GIS visualization has emerged intrinsic to representation and is not without challenges.

9.3.1.5 GIS and Society

Geographic information is no longer the mere concern of academics and policy makers but in a big way geographic information is a concern of people themselves and in so much it is a major concern of big businesses. In particular, in the post-fordist era, with flexible production system becoming the rule, digitized geographic information system has come to play a big role in the success of corporations. We might remind ourselves that the GIS enterprise with the ever growing need for data acquisition and dissemination, software development and applications is a multi-billion-dollar industry even by conservative assessments (Goodchild and Haining 2004). In turn, this has meant tremendous pressure on existing social, legal and institutional arrangements (Curry 1998; Hoeschele 2000).

The way GIS is constituted in most places makes it difficult for a lay person to critically engage in spatial planning processes. Besides, because GIS has become so visible in terms of its planning and development relevance in society, it has also acquired through its applications power, wealth and influence of special interests. Thus, with the potential for GIS applications to be appropriated to solve planning and development issues, the challenge then is how the technology can be utilized appropriately and in a way that minimizes its abuse and increases its efficiency (Hoeschele 2000). In most peripheral areas of both developed and developing countries GIS as a technology remains inaccessible to groups and individuals thus prohibiting constructive critical engagement of the community in the complexities of GIS applications to planning and development (Obermeyer 1998; Rambaldi et al. 2006). In fact, it is this empowering ability of GIS that has made Terry Jordan's claim of 'GIS being no more than a mere tool' redundant (Jordan 1988). As a result the last decade in particular has seen the emergence of 'public participation GIS or PPGIS' (Rambaldi et al. 2006).

The challenges that lie in PPGIS's practice in peripheral areas of developing countries emerge from the need to integrate low and high spatial information management applications. Particularly, in the rural contexts, PPGIS is practiced essentially through non-governmental organizations (NGOs) and community-based organizations (CBOs) as well as a wide array of development personnel and advocates who work as intermediaries between people and the local government bureaucracies (Rambaldi et al. 2006). Since PPGIS can only be effective if they tend to focus on processes by which outcomes are attained, the challenge often remains in how to obfuscate unequal and superficial participation by eliminating inequalities. While GIS and the attendant possibilities that PPGIS creates has been hailed as democratizing and yet most developing countries systems particularly in the Sub-Saharan Africa where governments are inherently authoritarian, thereby creating contradictory scenarios and posing challenges for GIS to permeate the current debate on transfer and applications of the technology to indigenous communities (Kyem 2000). To add to these challenges, most similar cases are unpublished.

On the other hand, PPGIS in the peripheral areas of the developed countries have evolved in response to the challenge of and the need to make GIS and its intersection with spatial planning process accessible to those whom it affects most, be it management of urban problems associated with the inner-cities or planning and developmental issues of indigenous communities where technical competency and cost have been barriers to GIS implementation, PPGIS applications in non-urban spaces of the developed countries have occurred in partnership between communities, universities, grassroots social organizations and internet-based initiatives of local and state government agencies (Ghose 2001). In the realm of PPGIS, Ghana's experience is contrary to popular expectations. The inflexible bureaucratic organizations there are not willing to let the power relations shift in favor of marginalized groups in society in the process of public participation and empowerment (Kyem 2001). Conflict of interest can pose another challenge in any participatory exercise. Ghana's experience and the attendant challenges can be seen extrapolated in other places such as North-west Cambodia where GIS and participatory methods have been integrated to assess risk to local communities from landmines and to develop priorities for landmine clearance. However, the challenge once again remains political since this part of Cambodia is still emerging from military conflict (Williams and Dunn 2003).

In a related concern, GIS applications in the developed world in particular have thrown up the issue of 'privacy law'. The challenge therein is that while GIS applications in the form of closed circuit television, radar-imaging, and other technologies connected with surveillance has caused people the concern in the developed world about the excessive penetration of university, science and the private sector into the privacy of individuals, particularly since corporations use GIS applications to understand shopping behavior, credit card debts, viewing habits etc. (Curry 1998).

9.3.2 Practical Challenges

This section presents several practical challenges in the domain of planning and development as is experienced through GIS applications and its integration with socio-economic and demographic statistics.

9.3.2.1 Infrastructure, Financing and Technology Transfer

The widespread use of GIS in peripheral areas of developing countries begs the need for installing a viable infrastructure and information systems tool in place. While the rate of adoption of GIS applications has increased rapidly in the last couple of decades, there are still considerable challenges in the implementation of GIS. In most developing countries the establishment of a spatial technology infrastructure is dependent on external funding or temporary support received through aid projects often introduced with a view to demonstrate the use and need of technology. Thus, when the financial support ends so ends the licensing of software and the motivation to maintain the initial efforts also disappears. While, some developing countries such as China, India, Malaysia, Thailand, Indonesia have committed their financial and human capital resources for appropriate infrastructure building of a high level of spatial information technology, many other countries in Africa, Latin America and South East Asia still continue to be challenged by lack of financial investment to acquire GIS technology, computing skills and inadequate availability of data. External aid attempts have provided limited training to some selected individuals abroad with the hope that when the trained personnel return they would diffuse the technology at home. Thus, the challenge in expanding GIS applications in planning and development has been marred by motivational, institutional and political constraints which are far less easy to overcome.

9.3.2.2 Development

A review of articles from a wide variety of journals such as Journal of Geographical Systems, Transactions in GIS, URISA, International Journal of Geographical Information Systems, Cartographica, Cartography and Geographic Information System, and mainstream journals in the discipline of geography such as Annals of the Association of American Geographers, The Professional Geographer and The Canadian Geographer shows that substantial attention has been paid to challenges emerging from the application of GIS in the domain of development, particularly in the case of peripheral areas of the developing countries.

In particular, GIS applications find penetration in the area of public health in both developing and developed countries though the nature of the challenge differs in both cases. In the case of the developed countries the challenge remains in the domain of spatial database consolidation, data warehousing, while protecting concerns arising out of confidentiality and privacy of health data (Cromley and McLafferty 2002). On the other hand, while GIS and Remote Sensing have together been used in many peripheral areas of developing countries such as Indonesia, Brazil, Nigeria and Uganda among many others to identify and address health issues. In Nigeria geostatistics and GIS were together used with remotely sensed environmental data to understand the distribution of urinary schistosomiasis or 'blood in urine' in Ogun state of Nigeria (Ekpo et al. 2008). In Uganda's Gulu region geostatistics such as Moran I Statistics, remotely sensed data (including DEM) along with data on geo-referenced aquatic habitats in various internally displaced camps (IDP) were used to model relationship between culex and anopheles aquatic habitats (Jacob et al. 2008). In a yet another case of community health assessment effort in a remote island village of Indonesia, efforts were made to identify and address health issues through socio-economic data, disease surveillance, health care utilization, environmental data and health resource allocation. However, the objective of analyzing data from disparate sources was possible through support from the University of Pittsburgh which designed software called SOVAT (Parmanto et al. 2008). Similarly, in a yet another attempt to construct a complex simulation model in network epidemiology in Brazil through software 'Epigrass' analysis made use of large geo-referenced databases (Coelho et al. 2008). A common challenge that was cited in each one of these cases was the difficulty in obtaining spatial databases particularly in the case of Uganda and Nigeria while in the case of Indonesia the challenge was the high cost of computing resources and the lack of computing skills necessary to support such an assessment.

Within the realm of development planning the significance of sustainable land use need not be overemphasized. In an example from India, attempts to alleviate land degradation through the participation of local communities with the use of GIS based applications still has to be within the government's overarching control (Puri and Sahay 2003). While landscape studies have been carried out in North America and Europe with a view to understand changes in land use and land cover, such exercises are conspicuously absent from the developing countries. Attempts to use GIS approaches to understand landscape change in Egypt have met with frustrations for lack of longitudinal data and limited large-scale data availability (Stewart 2001).

9.3.2.3 Municipal Planning and Infrastructure

In recent years municipal planning and infrastructure management is a domain where GIS applications have been widely adopted particularly in the developing countries. In India, use of GIS has been introduced for a wide variety of purposes ranging from surveying and mapping slums in Pune-Sangli region with a view to integrate them into the city-wide urban planning process to developing a solid waste management system using GIS (Joshi et al. 2002; Ogra 2004). The Pune slum projects faced the challenge of being stalled because of strained relations between administrators and politicians. On the other hand, the problem with the solid waste management based spatial planning project was challenged by the data being too spread and isolated for use in planning, management and decision making.

9.4 Conclusions

This chapter addressed the limitations of GIS applications in both methodological and practical terms for existing governments and other stakeholders at all scales in both developed and developing countries. Even while there is awareness about the applications of GIS and its potentials it seems like the journey is going to be long and arduous before developing countries too become self reliant in terms of spatial decision support infrastructure. Clearly the challenge for both developed and developing countries vary. In a sense developed countries need to hone their capacity in terms of further integrating GIS into their system, perform cost-benefit analysis, manage implementation and operations in a dynamic environment, while the developed countries need to constantly find ways to include GIS applications both horizontally and vertically more entrenched in the planning and development domain through demonstration of widespread application potentials, particularly, in critical cost saving areas such as urban planning and management, poverty eradication and livelihood creation and public health and natural resource management.

GIS is spreading rapidly in the peripheral regions of the world creating enormous opportunities for local regional and national governments. The challenge for those in planning and governance lies in the need to decide to what extent GIS should be incorporated into their day-to-day operations and how to go about implementing them. The fact that there is disparity between infrastructures of wealthy and poor countries could be overcome by developing a network infrastructure. There is tremendous hope for sharing knowledge, transfer of technology and convergence of interest leading to financing and construction of spatial and digital information infrastructure in peripheral regions.

References

Agnew, J. (1993) Representing Space: Space, Scale and Culture in Social Science. In James Duncan and David Ley (eds.) Place/Culture/Representation. London: Routledge.

Anselin, L. (2000) GIS, Spatial Econometrics and Social Science Research, *Journal of Geographical Systems*, vol. 2, pp. 11–15

Brail, R. K. and Klosterman, R. E. (2001) Planning Support Systems: Integrating Geographic Information Systems, Models and Visualization Tools, Redlands, CA: ESRI.

Coelho, F. C., Cruz, O. G. and Codeco, C. T. (2008) Epigrass: A Tool to Study Disease spread in Complex Networks, http://www.scfbm.org/contents/3/1/3

Cox, K. (1998) Spaces of Dependence, Spaces of Engagement and the Politics of Scale, *Political Geography*, vol. 17, pp. 1–24.

Cromley, E. K. and McLafferty, S. L. (eds.) (2002) *GIS and Public Health*. New York: Guilford Press.

Csillag, F. and Boots, B. (2005) A Framework for Statistical Inferential Decisions in Spatial Pattern Analysis, *Canadian Geographer*, vol. 49, no. 2, pp. 172–179.

Curry, M. R. (1995) GIS and the Inevitability of Ethical Inconsistency. In John Pickles (eds.) *Ground Truth: The Social Implications of Geographic Information Systems*. New York: Guilford.

Curry, M. R. (1998) *Digital Places: Living with Geographic Information Technologies*. London: Routledge

Drummond, W. J. (1995) Address Matching: GIS Technology for Mapping Human Activity Patterns, *Journal of the American Planning Association*, vol. 61, no. 2, pp. 240–251.

Ekpo, U. F. et al. (2008) Geographical Information System and Predictive Risk Maps of Urinary Schistosomiasis in Ogun State, Nigeria, http://www.biomedcentral.com/

Garner, B. J. and Zhou, Q. (1991) GIS Education and Training: An Australian Perspective, *Published in the Proceedings of GIS/LIS*, pp. 388–398.

Gatrell, J. (1999) Re-Thinking Economic Development in Peripheral Regions, *The Social Science Journal*, vol. 36, no. 4, pp. 623–639.

Ghose, R. (2001) Use of Information Technology for Community Empowerment: Transforming Geographic Information Systems into Community Information Systems, *Transactions in GIS*, vol. 5, no. 2, pp. 141–163.

Goodchild, M. F. and Haining, R. P. (2004) GIS and spatial data analysis: Converging perspectives. *Regional Science*, vol. 83, pp. 363–385.

Hall, G. B. (1999) GIS Education and Infrastructure: Challenges and Problems in Emerging Countries, *Transactions in GIS*, vol.3, no. 4, pp. 311–317.

Hastings, D. A. and Clark, D. M. (1991) GIS in Africa: Problems, Challenges and Opportunities for co-operation, *International Journal of Geographical Information Science*, vol. 5, no. 1, pp. 29–39.

Herod, A. (1991) The Production of scale in the United States labor Relations, *Area*, vol. 23, pp. 82–88.

Hoeschele, W. (2000) Geographic Information Engineering and Social Ground Truth in Attappadi, Kerala State, India, *Annals of the Association of American Geographers*, vol. 90, no. 2, pp. 293–321.

Jacob, B. G. et al. (2008) Hydrological Modeling of Geophysical Parameters of arboviral and Protozoan disease vectors in Internally Displaced People Camps in Gulu, Uganda, *International Journal of Health Geographics*, vol. 7, no. 11.

Jankowski, P. and Nyerges, T. (2001) GIS-Supported Collaborative Decision Making: Results of An Experiment, *Annals of the Association of American Geographers*, vol. 91, no. 1, pp. 48–70.

Jonas, A. (1994) The Scale Politics of Spatiality, *Environment and Planning D: Society and Space*, vol. 12, pp. 257–264.

Jordan, T. (1988) The Intellectual Core: President's Column, *AAG Newsletter*, vol. 23, no. 1.

Joshi, P., Sen, S. and Hobson, J. (2002) Experiences with Surveying and Mapping Pune and Sangli Slums on a Geographical Information System, *Environment and Urbanization*, vol. 14, no. 2, pp. 225–240.

Karikari, I., Stillwell, J. and Carver, S. (2005) The Application of GIS in the Land Sectors of a Developing Country: Challenges facing Land Administrators in Ghana, *International Journal of Geographical Information Science*, vol. 19, no. 3, pp. 343–362.

Kwan, M. (2002) Feminist Visualization: Re-envisioning GIS as a Method in Feminist Geographic Research, *Annals of the Association of American Geographers*, vol. 92, no. 4, pp. 645–661.

Kyem, P. A. K. (2000) Embedding GIS Applications into Resource Management and Planning Activities of Local and Indigenous Communities: A Desirable Enterprise? *Journal of Planning, Education and research*, vol. 20, pp. 176–186.

Kyem, P. A. K. (2001) Power, Participation, and Inflexible Institutions: An Examination of the Challenges to Community Empowerment in Participatory GIS Applications, *Cartographica*, vol. 38, no. 3 & 4, pp. 5–17.

Lam, N. and Quattrochi, D. A. (1992) On the Issues of Scale, Resolution, and Fractal Analysis in the Mapping Sciences, *The Professional Geographer*, vol. 44, pp. 88–98.

Lloyd, R., Hodgson, M. E. and Stokes, A. (2002) Visual Categorization with Aerial Photographs, *Annals of the Association of American Geographers*, vol. 92, no. 2, pp. 241–266.

Maguire, D. J., Goodchild, M. F. and Rhind, D. W. (eds.) (1991) Geographical Information Systems. Vol. 2. New York: John Wiley & Sons.

Martin, D. (2003) Reconstructing Social GIS, *Transactions in GIS* vol.7, no. 3, pp. 305–307.

Massey, D. (1994) Space, Place and Gender. Minneapolis: University of Minnesota Press.

Nijkamp, P. and Scholten, H. J. (1993) Spatial Information Systems: Design, Modeling and Use in Planning, *International Journal of Geographical Information Systems*, vol. 7, pp. 85–96.

Nkambwe, M. (2001) Land Information Systems Development and the National Spatial Data Infrastructure in Botswana, *The International Archives of the Photogrammetry, Remote Sensing and Spatial Information Sciences*, vol. 34, no. 6.

Obermeyer, N. (1998) The Evolution of Public Participation GIS, *Cartography and Geographic Information System*, vol. 25, pp. 65–66.

Ogra, A. (2004) Logistics Management and Spatial Planning for Solid Waste Management System using Geographic Information System, http://www.gisdevelopment.net

Parmanto, B. et al. (2008) Spatial and Multidimensional Visualization of Indonesia's village health statistics, *International Journal of Health Geographics*, vol. 7, no. 30.

Puri, S. K. and Sahay, S. (2003) Participation Through Communicative Action: A Case Study of GIS for Addressing Land/Water Development in India, *Information Technology for Development*, vol. 10, pp. 179–199.

Quattrochi, D. A. and Goodchild, M. F. (eds.) (1997) Scale in remote Sensing and GIS. New York: CRC Publishers.

Rambaldi, G., Kyem, P.A.K., McCall, M. and Weiner, D. (2006) Participatory, Spatial Information Management and Communication in Developing Countries, *The Electronic Journal on Information Systems in Developing Countries*, vol. 25, no. 1, pp. 1–9.

Schuurman, N. (2004) GIS: A Short Introduction. Malden, MA: Blackwell.

Schuurman, N. (2006) Formalization Matters: Critical GIS and Ontology Research, *Annals of the Association of American Geographers*, vol. 96, no. 4, pp. 726–739.

Smith, D. M. (1992) Contours of a Spatialized Politics: Homeless Vehicles and the Production of geographic scale, *Social Text*, vol. 33, pp. 55–81.

Steinmann, R., Krek, A. and Blaschke, T. (2004) Analysis of Online Public Participatory GIS Applications with Respect to the Differences between the US and Europe, *Paper published in the proceedings of Urban Data Management Symposium'04, Chioggia, Italy*.

Stewart, D. J. (2001) New Tricks with old Maps: Urban Landscape change, GIS, and Historic Preservation in the Less Developed World, *The Professional Geographer*, vol. 53, no. 3, pp. 361–373.

Sui, D. Z. (2004) GIS, Cartography, and the 'Third Culture': Geographic Imaginations in the Computer Age, *The Professional Geographer*, vol. 56, no. 1, pp. 62–72.

Williams, C. and Dunn, C. E. (2003) GIS in Participatory Research: Assessing the Impact of Landmines on Communities in North-West Cambodia, *Transactions in GIS*, vol. 7, no. 3. pp. 393–410

Yapa, L. S. (1991) Is GIS appropriate technology? *International Journal of Geographical Information Systems* vol. 5, pp. 41–58.

Chapter 10
Geospatial Technologies for Surveillance of Heat Related Health Disasters

Daniel P. Johnson

Abstract Heat Related Health Disasters (HRHD) and Extreme Heat Events (EHE) are currently a major public health and climate change concern. EHEs are the number one cause of death in relation to environmental disasters; precipitated as an HRHD. Thought to exacerbate this phenomenon in urban settings is the Urban Heat Island (UHI) effect. Moreover, over 50% of the current worldwide population resides in an urban setting. Therefore the need of a system to specify spatially the areas of increased risk due to an EHE is apparent. The conceptualization of such a system is presented in a parsimonious fashion involving the description of geostatistical methods and thermal remote sensing platforms. Socioeconomic indicators of risk, to extreme heat, are discussed with how they potentially blend with neighborhood level thermal characteristics obtained from remotely sensed assets. Modeling such relationships is discussed with logistic regression and artificial neural networks. The primary proposed outputs are cartographic products elucidating risk from HRHDs. Such geospatial techniques have intrinsic abilities to both plan for and mitigate urban disasters. This conceptualization should assist medical geographers, public health practitioners and researchers in planning for the surveillance of HRHDs.

Keywords Medical geography · GIS · Heat related deaths · Spatial analysis

10.1 Introduction

Geospatial technologies have been in used in health related applications for several decades. However, their full potential in the surveillance of environmentally related health disasters has yet to be realized. Currently, humankind is faced with many types of potential environmental health disasters. Many of these putative hazards are infectious diseases such as malaria, schistosomiasis, dengue fever and AIDS (Lorentzen et al. 2008; Remais et al. 2008; Schaeffer et al. 2008; Vezzani and

D.P. Johnson (✉)
Indiana University Purdue University at Indianapolis, Indianapolis, IN, USA
e-mail: dpjohnso@iupui.edu

J.D. Gatrell, R.R. Jensen (eds.), *Planning and Socioeconomic Applications*, Geotechnologies and the Environment 1, DOI 10.1007/978-1-4020-9642-6_10,

Carbajo 2008). Moreover, many potential disasters involve suspected scenarios of climate change which likely lead to an increase in extreme weather events (Katz and Brown 1994). Disasters such as tropical cyclones, extreme heat events (EHE) and drought can lead to an immediate health disaster and provide putative components which exhibit latency; thus prolonging the detrimental health impact. Detailing such scenarios involve the work of geographers, disaster epidemiologists, and other researchers with medical training.

Evidence suggests that climate change is a natural process that is being anthropogenically exacerbated by numerous inputs (von Storch and Stehr 2006). An immediate and primary health concern with current climate models is the expected increase in duration and frequency in EHEs (Koffi and Koffi 2008). Heat related mortality is the number one weather related killer in North America; killing more people than floods, lightening, hurricanes, hypothermia and tornados combined. This statistic is likely the same for regions outside of North America and is suspected to be drastically underreported. For example, in the United States, many people die during EHEs from conditions that are exacerbated by extreme heat, not hyperthermia itself (Mastrangelo et al. 2006). The medical examiner conducting the autopsy may or may not record the death as being induced by heat.

There is a lack of surveillance activity for these types of events throughout the world. Some cities in the United States currently use the Heat Health Watch Warning System (HWWS) developed by Lawrence Kalkstein (Kalkstein 1991; Kalkstein et al. 1996). This system provides a heat alert for an entire city or metropolitan area. During such an emergency, health personnel as well as the general public are encouraged to "check in" on elderly people and watch for signs of persons succumbing to heat stress. However, most cities are inadequately informed on where the most vulnerable people live and which areas of the city are most at risk during EHEs. Geospatial technologies provide a valuable resource to assist in modeling this type of disaster phenomenon and identifying areas most at risk. The HWWS is limited by its ability to identify this intra-urban variation in risk. Therefore a system utilizing a range of geospatial technologies will likely assist in identifying this risk as well as mitigate the disaster.

Following is a discussion on geospatial technologies as they relate to HRHDs. Provided is a parsimonious outline of how such technologies might be used to assist in identifying areas of risk as well as mitigate the health effects of an EHE. It is hoped that such delineation may assist public health professionals in conceptualizing how to utilize geospatial methods for disaster response and preparation.

10.2 Heat Related Health Disasters (HRHD)

HRHDs currently lead the list of disasters that impact human life across the globe. They not only lead the list of weather related phenomena that kill but also lead in their relative frequency of occurrence. Such aspects of extreme heat are typically not known to the general public (Kalkstein and Sheridan 2007; Sheridan 2007; Semenza et al. 2008). This leads to the unfortunate occurrence of heat being defined as a "silent killer" (Klinenberg 2002). For example, one cannot look outside and see the heat wave as one can an approaching storm. HRHDs exacerbate medical symptoms across the spectrum of human health (Semenza et al. 1999). Moreover, it is likely that mortality associated with such events is drastically underreported due to the lack of a universal standard in how a medical examiner reports heat related causes of death (Mastrangelo et al. 2006).

10.2.1 Heat Events Defined

There is no universally accepted definition of what parameters correspond to a heat wave or heat event. Underscoring this is the different nomenclature used to describe such events. For example, in the state of California a heat wave is typically called a heat storm by the California Energy Commission. The National Weather Service further defines a heat wave as "A period of at least 48 hours during which neither the overnight low nor the daytime heat index falls below the NWS heat stress thresholds of 80F and 105F respectively" (NOAA 2008). Additionally there are several variations of this definition ranging in intensity from an "intense heat wave" to a "hot spell" (Robinson 2001). Typically for such an event to lead to mortality of vulnerable groups the event needs to last approximately 2–4 days (McGeehin and Mirabelli 2001). For purposes of discussion here an extreme heat event (EHE) is the universal name given to extended weather related heat conditions and a heat related health disaster (HRHD) is defined as an EHE leading to excessive mortality.

10.2.2 Vulnerability

There is limited knowledge of where the groups most vulnerable to extreme heat reside within urban centers. The identification of such vulnerable neighborhoods is necessary in the development of spatially-specific intervention strategies. The most vulnerable groups, not the location of them, to extreme heat and other urban disasters have been well studied (Cutter 2003). Persons living in poverty, with less than a high school education, belonging to a minority group, and the very young and the very old are disproportionately affected by EHEs (Smoyer et al. 2000). The very old are especially at risk due to a loss of social contacts and the physical impact increased age has on thermoregulation (Klinenberg 2001; Basu and Samet 2002). Heat exacerbates many medical conditions and neighborhood-level thermal characteristics are an important factor in heat-related mortality (Smoyer 1998; Harlan et al. 2006). Additionally some research now suggest that populations are adapting to some climatic fluctuations, thus leading to a decrease in mortality from EHEs (Fouillet et al. 2008). However, more work needs to be done in this area and any adaptation is likely spatially and temporally conditional.

It has been suggested in numerous studies that the urban heat island (UHI) effect may play an exacerbating role in heat-related death (Buechley et al. 1972; Ebi and

O'Neill 2008). The UHI is the observed difference between the rural and urban temperatures, observable as increased temperatures in the urban center (Saaroni et al. 2000). This phenomenon has been observed since the late 1800s. Thermal remote sensing instruments offer a way of delineating the surface UHI which is very dependent on the land cover and land use characteristics of the respective city's built environment (Voogt and Oke 2003; Rosenzweig et al. 2005). Potentially coupling thermal data with socioeconomic indicators of vulnerability will lead to detailed models of risk illuminating intra-urban variability. Explaining this variability will lead to increased knowledge of risk within urban centers.

10.2.3 Historically Significant HRHDs

St. Louis, MO has witnessed significant HRHDs in the past 50 years. In July of 1980 the incidence of all cause mortality increased 57% over the previous two years (Jones et al. 1982). Additionally, the rate of death during this event was 26.5 per 100,000 people. The most vulnerable during this event were the urban poor and the elderly (Jones et al. 1982; Smoyer 1998). Another HRHD occurred in St. Louis in the summer of 1995 which resulted in high rates of mortality. Additionally, it was demonstrated by Smoyer, comparing the 1980 to the 1995 HRHD, that mortality risk from extreme heat events has likely increased, within St. Louis, even with increased utilization of residential air conditioning (Smoyer 1998).

Philadelphia, PA has also seen several HRHDs during the past few decades. An event in Philadelphia in 1993 claimed over 116 lives (CDC 1994; Mirchandani et al. 1996). Additionally, in May of 2008 seventeen people succumbed to extreme heat in Philadelphia (Newbern 2008). Philadelphia is particularly vulnerable to HRHDs due to several factors. The first is the numerous concentrations of older stock homes within the city. These homes are built to withstand extremes in winter temperatures and essentially serve as a "brick oven" during EHEs. Additionally, these older homes contain older electrical components which are not able to support the power needs of air conditioning systems. These two factors alone add to an increase in risk for those living in this environment; typically the urban poor which are unable to move or upgrade electrical capacity.

Perhaps the most significant HRHD in the United States is the one that occurred in Chicago, IL in July of 1995 (Klinenberg 2001). This particular event claimed 739 lives and brought the realization of HRHDs to the forefront of public perception. From July 11th to July 17th temperatures ranged from a high of 106 F and rarely fell below 80 F in the evening. This led to increases in mortality due to exposed populations rarely receiving adequate relief. This event also led to a significant increase in hospital admissions for a variety of conditions (Semenza et al. 1999). Additionally, it has been shown that this was a policy disaster inside a meteorological one (Klinenberg 2001, 2002). Many of the public health personnel within the city were not aware that heat was indeed killing exposed vulnerable persons. The hospitals and morgues quickly filled with those succumbing to the extreme in high temperature. After the 1995 event the city of Chicago instituted cooling centers within the city to aid in relief during heat events. These cooling centers are public spaces which have adequate air conditioning as well as places for persons to sit and rest during their stay.

The heat event that occurred in Europe in 2003 further underscored the health impact of climate changes and associated mortality. This event caused over 35,000 deaths in Europe, almost 15,000 in France alone (Beniston 2004). Occurring from August 1st to August 20th temperatures reached as high as 107 F and rarely dropped below 80 F in the evening. Additionally, many governments did not have a plan for a catastrophic heat event, thus leading to an increase in mortality from inadequate response. The majority of deaths occurring in Europe were elderly persons in hospice and nursing homes (Fouillet et al. 2006; Kovats and Ebi 2006). These individuals were at increased risk due to the inadequacy of air conditioning and the older buildings in which they were institutionalized.

10.3 Methods of Surveillance

10.3.1 Current Surveillance and Warning Systems

Currently the most widely-used method of alert for extreme heat is the Heat Watch/Warning System (HWWS) developed by Lawrence Kalkstein (Sheridan and Kalkstein 2004). This system is primarily based on synoptic models that forecast the weather components necessary to prompt an extreme heat alert. This type of system has viability as it fits in very well with the current state of how the National Weather Service (NWS) issues warnings and/or watches for extreme weather events. As of this writing the system is now established in 20 cities in the United States and numerous cities internationally. The NWS also has a heat alert system which is issued for broad ranging areas (larger scale than the city) when meteorological conditions are expected to result in a heat event (Robinson 2001). These systems are primarily involved in issuing of heat alerts for areas; although Kalkstein's model is more specific to a given area.

Developed from a concern the heat index guide used by the NWS was insufficient to characterize risk, the HWWS was developed by the Center for Climatic Research (Kalkstein et al. 1996). Primarily this interest was rooted in the arbitrariness of the heat index values; not being specifically defined for certain locations. For example, a heat index in Dallas, TX of 105 F would not be as substantial as one of 105 F for Chicago, IL. Residents of Dallas are likely more accustomed to such values than residents of Chicago. It was thought that additional meteorological variables should be included as well as linkages to health data (Kalkstein 1991).

The HWWS is based on a daily classification of numerous meteorological conditions for a specific location (Kalkstein et al. 1996; Sheridan and Kalkstein 2004). Measurement such as temperature, dew point, wind speed/direction, and cloud cover are acquired throughout the day. These measurements are typically recorded by the NWS office locations in each city. There is a daily assignment of an air mass type which is correlated to days with significant increases in death. From these calculations probabilities can be developed to determine which conditions (air masses) are expected to contribute to increased heat related mortality. This system also intrinsically provides the ability to account for decreases in mortality as a result of population adaptation. The HWWS was first implemented in Philadelphia PA; a site of numerous heat events and a high count of vulnerable persons.

Ebi developed a performance metric of the HWWS for Philadelphia following implantation of the system (Ebi et al. 2003). The development of this metric involved a multiple linear regression analysis of the following variables: daily temperature readings, heat wave duration in days, time of season, air mass origin and issuance/non-issuance of an extreme heat alert. It was found that 2.6 lives per day (during a heat event) were saved due to the issuance of a heat alert. Using the value of statistical life (VSL) as measured from the EPA ($4 million for persons 65 and over) it was estimated that over $468 million dollars was saved in the three-year period of the study. Moreover, this VSL measure drastically exceeds any cost of implementing the system.

The HWWS, although successful and imperative, is limited by its inability to identify intra-urban variations in risk based on contributing factors such as socio-economic status and land cover/land use patterns. Additionally, it focuses on an entire city and does not contain a component that will allow emergency personnel to target areas of the city. In order to promote effective spatially targeted intervention during such a disaster finer scale modeling of the phenomena is needed. Such a system should involve a component which begins with the description of the distribution of vulnerable populations. Further it should utilize information assets available from the numerous remote sensing platforms; including thermal components. Such assimilation would complement the HWWS in a way that would foster intra-urban measures of risk. This would enable effective mitigation during a HRHD and allow for planning prior to such events. Additionally, the development of such a system should follow known methods of geostatistics and digital image processing.

10.3.2 Geostatistical Analysis

Geostatistical techniques have been used to evaluate the distribution of health related phenomena (Kulldorff 2001; Jemal et al. 2002; Kulldorff et al. 2003). For such an approach to be effective in the surveillance of HRHDs it would first be necessary to collect data from a census for determining the numbers of at-risk persons living within a given area. These data could be analyzed in conjunction with mortality data for correlation with areas of suspected risk. This descriptive analysis would be well suited for preliminary investigations into disaster preparation. For example, if a clear picture of vulnerable population distributions is acquired before a particular event the impact on certain areas may be lessened.

The SDE, one of the simplest geostatistical measures, was first introduced in 1925 by a sociologist interested in geographic concentration (Furfey 1927). Since that time it has been employed in numerous studies as a candidate for spatially descriptive exploration (Levine 1995; Wong 1999, 2003, 2005). This method generates an ellipse which encompasses a z value of the point distribution. Additionally, the SDE depicts directional bias present in a set of points (Wong 2005). However, this directional bias sometimes aligns fairly well with the study area under investigation. For example if the study area lies in a strict southwest to northeast pattern, the ellipse will likely take on this angle as well. The calculation of the SDE is simplistic and straightforward. Initially it is important to calculate the mean center of the set of points; another useful measure of centrality. Once the mean center is calculated each point is then transformed into a different metric space referenced from the calculated mean center.

Figure 10.1 presents a typical SDE and some of the accompanying measures that are useful for the comparison of distributions. This measure differs from a standard distance circle (SDC) in that an SDC does not show the directional bias but just the standard distance present in a spatial distribution. This aspect is driven from the orientation measure which shows the deviation from true north in an angular metric. Moreover, this angular measurement is present in the X and Y standard distances. Eccentricity is a measure of the skewness of the distribution. A SDC would have a skewness of 1 and the SDE present in Fig. 10.1 has a skewness of 0.5. This measure is very useful in determining the polarity of the point distribution. For instance, a SDE with an eccentricity which is very low (close to 0.1) would suggest that the spatial phenomena investigated would demonstrate high polarity;

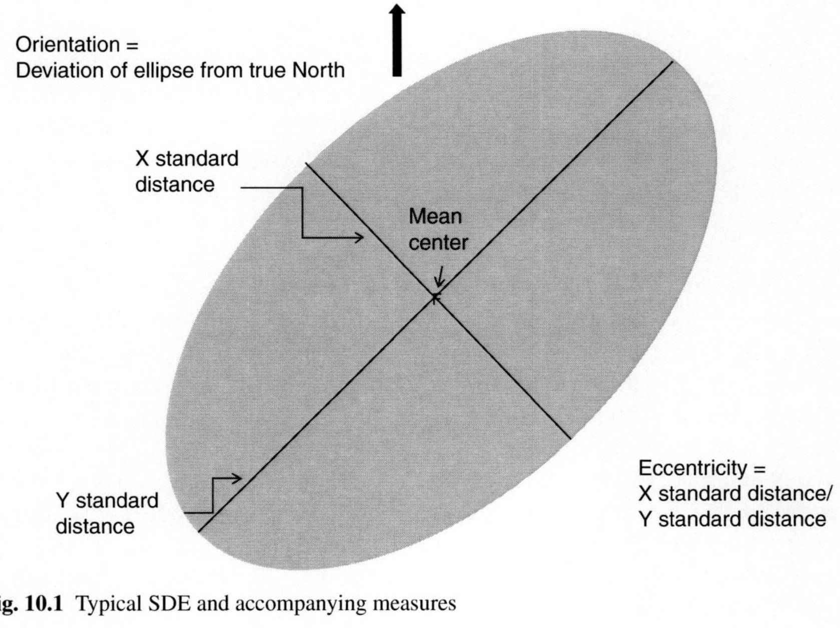

Fig. 10.1 Typical SDE and accompanying measures

all the distribution would be at both ends of the Y standard distance with little in the middle. These measures are spatially equated to aspatial measures of central tendency.

The use of the SDE leads to a descriptive measure of the distribution of the population of points under consideration. Distributions can be compared to one another to measure levels of similarity. For example, if the distribution of death from extreme heat corresponds highly to the distribution of Hispanic population there is likely more at play than simple spatial congruency. Further examination of this might reveal that the thermal properties within the city correspond highly with death and Hispanic population. This could be demonstration that the Hispanic population residing in areas of the city which promote thermal storage in the environment. Determining such thermal properties within the city is the focus of the next section.

Other useful methods include polygon pattern descriptors which focus on the level of clustering within the study area. These methods include global descriptors such as Moran's I and Geary's C measures of spatial autocorrelation (Unwin 1996; Zhang and Lin 2008). However, they also include measures of local autocorrelation such as the G-statistic developed by Getis and Ord (1992; Ord and Getis 1995). Such measures seek to elucidate the complex relationships of spatial phenomena and how they relate one another spatially, specifically within enumerated units. This is a more complicated method than employing the SDE and leads to quantitative measures of clustering and dispersion. Care must be taken when using these methods as they are scale dependent.

Spatial autocorrelation describes the tendency of spatial phenomena in close proximity to one another to be similar in value. For example, if a study area consists of 12 census tracts and 6 of them near the center of the study area have similar values then one would expect a strong positive value of spatial autocorrelation. Moran's I and Geary's C provide a value for the study area describing this spatial relationship. Values for Moran's I theoretically range from 1 to -1. Positive values exhibit positive spatial autocorrelation and negative values negative spatial autocorrelation. Geary's C is inversely related to Moran's I (Unwin 1996). Values for this ratio range from 0 to 2. A value of 0 indicates perfect positive spatial autocorrelation and a value of 2 indicates perfect negative spatial autocorrelation. Both measures are related to one another and it is recommended that both measures be calculated for a given study.

There are limitations to the use of Moran's I and Geary's C. The most substantial limitation when dealing with medical phenomena is that both measures are global in nature. Thus, the calculation of spatial autocorrelation is done for the entire study area. For example, in the study area consisting of 12 census tracts, we might receive a value for Moran's I of 0.7. This indicates strong spatial autocorrelation but we are not provided with information as to where this cluster of similarity is. This is relatively unimportant when dealing with 12 census tracts but if the study area were extended to 500 census tracts the location of the similarity may be more difficult to define. Such information is typically vital when dealing with health and risk. Measures dealing with localized measures of spatial autocorrelation are more appropriate for such investigation.

One of the more powerful measures of local autocorrelation is the G statistic developed by Getis and Ord (1992). By incorporating this measure a G statistic is developed for each polygon area under consideration by incorporating a spatial lag component. The spatial lag factor effectively consults the Euclidean distance from polygon to polygon. Utilizing this measure results in a statistic that describes how similar spatial components are to one another based on distance. Values of high next to other high values indicate strongly positive spatial autocorrelation and low next to low strongly negative; values in between range from moderately positive to moderately negative (Ord and Getis 1995). The utilization of such a measure in health or risk related applications would lead to indications of local scale variations in clustering. Such measures are vitally important when dealing with the development of spatially specific intervention scenarios and disaster mitigation.

10.4 Thermal Remote Sensing

Thermal remote sensing platforms offer a way of measuring the thermal inertia of objects on the Earth's surface. These platforms typically record energy in the thermal infrared wavelengths from 10.5 to 12.5 μm; as well as non-thermal wavelengths. These sensors can capture cloud temperatures as well as land surface temperatures useful for a number Earth resource applications. Primarily the consideration here is the usefulness of these systems to estimate land surface temperature (LST). Such systems have ability to determine the extent, magnitude and spatial variation of the surface UHI phenomena. This leads to consideration of neighborhood level variations in surface temperature which might lead to increased risk during EHEs (Wang et al. 2004). For example, if certain neighborhoods thought to be at risk due to socioeconomic conditions also present high intensity levels of LST then such neighborhoods are likely at an even greater risk during HRHDs.

10.4.1 MODIS

The MODerate resolution Imaging Spectroradiometer (MODIS) has been utilized for delineating the UHI and determining heat flux patterns for inclusion in Global Circulation/Climate Models (GCM) (Jin et al. 2007; Wang et al. 2007). This particular instrument has 36 bands between 0.405 μm and 14.385 μm. It contains two bands between 10.780 and 12.270 μm which can measure both surface and cloud temperature; with a spatial resolution of 1 km × 1 km. Since spatial and temporal resolution are typically inversely related MODIS can capture surface temperatures (LST) globally everyday.

MODIS LST data for both day and night can be collected for days corresponding to high temperatures and excessive deaths for the respective cities and their encompassing Metropolitan Statistical Area (MSA). This can provide indication to the extent to which vulnerable populations correspond to elevated surface temperatures. The 1 km spatial resolution of MODIS can be utilized to determine a broad area of risk within an MSA but is likely insufficient to determine intra-urban variability. Therefore, an instrument with a finer spatial resolution is likely needed to determine neighborhood level variations in LST.

10.4.2 Landsat

Band 6 of the Landsat TM instrument collects energy in the range of 10.4 to 12.5 μm. Additionally, Landsat TM data has been demonstrated to be effective in sensing the UHI (Aniello et al. 1995; Lo and Quattrochi 2003). The continuity of the Landsat program, allows for the acquisition of data pertaining to historic events. Moreover, the 90 m spatial resolution of the thermal band is appropriate for determining neighborhood level surface temperature variations. Landsat ETM+ also contains a thermal band within the same wavelength range as Landsat TM but provides an increased spatial resolution of 60 m. The increase in spatial resolution can lead to enhanced ability to explore intra-urban variation in surface temperature.

Landsat data does have limitation in regard to the radiometric and spectral resolution of the thermal band, leading to an error of ~5 K in LST estimation (Aniello et al. 1995). Additionally the data from both Landsat instruments is acquired every 16 days. This can lead to an inability to acquire an image during a heat event which might occur in between revisit dates. However, it is likely that it is not vital to collect the remotely sensed imagery during the EHE, although such a situation would be ideal. Since the UHI effect is driven by factors such as land use and land cover characteristics such components will likely not change in a short period of time. Therefore, collection of imagery very close to the event will likely be substantial enough for analysis provided there are similar surface conditions.

10.4.3 ASTER

The Advanced Spaceborne Thermal Emission and Reflection Radiometer (ASTER) is a variable temporal resolution sensor which acquires data in 5 thermal channels ranging from 8.125 μm to 11.65 μm. The spatial resolution of the instrument, for the thermal bands, is 90 m. This matches the Landsat TM resolution and is sufficient enough to capture intra-urban variations in the UHI (Tiangco et al. 2008).

ASTER is available as both surface kinetic temperature (SKT) and surface emissivity. Emissivity measures a materials ability to radiate absorbed energy and is related to surface temperature (Coll et al. 2007; Liu et al. 2007) Thus such measures might provide a more accurate description of the thermal load variability within an urban area. ASTER also provides a 1 C/K approximation of LST, a much more effective measure than Landsat even when corrected for emissivity. The temporal variability of the ASTER instrument also provides for the collection of data during the day and evening during. Such measures are vital to the exploration of surface temperature variations during an EHE which lead to increased risk due to environmental exposure.

10.5 Integration of the Technologies

Assimilating the technologies or methods of geostatistical analysis and remote sensing with health surveillance is not an easy task. However, it is necessary that such integration take place to both preserve public health and take advantage of the power inherent in geospatial technologies. The implementation of such systems is even more imperative in urban settings, where now over 50% of the world population reside. Following is a conceptualization of integrating such systems in anticipation of HRHDs in a city in the United States.

A likely first step is to collect data from the US Census Bureau pertaining to socioeconomic variables likely to indicate risk in urban disasters. Such variables include:

1. Hispanic population
2. Black population
3. Asian population
4. Native American population
5. Other race
6. Age 65 and over
7. Age 65 and over living below poverty
8. Age 5 and under
9. Persons below poverty
10. Low education (persons with less than high school education)
11. Building stock
12. Persons living alone

(Cutter 2003)

However, one is not limited with these variables and it is likely that an extended list may provide a more accurate or robust indication of risk. These variables should ideally be collected at either the census tract or block group level. Census tracts and block groups are good indicators of neighborhood level variations in risk or health (Smoyer 1998; Harlan et al. 2006). These variables, once identified, could be measured using Moran's I and Geary's C to determine global levels of autocorrelation within the city. For example, if one finds a particular vulnerable group highly clustered it may be at increased risk for heat related health effects, depending on the cluster location. Following the global measure one could utilize the G statistic to determine local scale variations in risk. This would provide cartographic outputs spatially delineating the clustered areas within the city. Additionally, one could convert the enumerated units to points and derive SDEs of each group (Wong 1999). These SDEs could then be analyzed using descriptive exploration via F and T tests to determine spatial similarity and or segregation between groups (Ned Levine 1995; Wong 1999, 2003, 2005). Descriptive analysis of vulnerable distributions within the city is vital in order to determine the level of dispersion or clustering of each group. For example, if a vulnerable group is spread throughout the city this may indicate a lower level of risk or a more difficult spatial intervention strategy; attempting to intervene within a dispersed population.

Incorporating the thermal remote sensing data is a logical extension to the socioeconomic analysis. Ideally by looking at local scale variations in LST derived from remotely sensed assets one can begin to assign LST values to each census tract or block group. If one were to do a multi-date analysis using Landsat ETM+ data, one would need to ensure that the imagery collected was relatively cloud free and that similar thermal conditions were present at the time of capture. Then geometric correction of the images would need to be performed to spatially match each image to a corrected spatial coordinate system. Then at-satellite-brightness temperature could be calculated from the thermal band (Aniello et al. 1995). At-satellite-brightness temperature is an estimation of LST but typically has a 10 F error of approximation when using Landsat data. Following this procedure one could assign temperatures to each enumerated unit and analyze them in conjunction with socioeconomic indicators. Such analysis could proceed via logistic regression, Bayesian Maximum Entropy (BME) or Artificial Neural Networks (ANN).

Logistic regression, or LOGIT modeling, could develop parsimonious models of risk and has been utilized in previous disease mapping studies (Glass et al. 1995). This type of modeling is used to determine the probability of a certain occurrence either numerically or categorically. The LOGIT model produces an odds ratio which can be used to determine the probability of an occurrence; in this case heat related mortality or illness. For example, based on socioeconomic conditions and surface thermal characteristics of a certain census tract there may be a 50% increase in the probability of death from extreme heat. In order to facilitate such analysis census tracts and block groups within each study area would be dichotomized based on certain variables. The remotely-sensed and socioeconomic variables would then be used to create the odds ratios as they relate to mortality within the respective urban area.

ANNs are nonlinear statistical modeling tools and have often been used for classification and pattern recognition in large datasets and often outperform other classification methods (Michie et al. 1994). ANNs have also been employed successfully in disease mapping and risk assessment (Kimes et al. 2004). The foundation of ANNs is that they mimic mammalian brain function and are able to "learn" patterns within a dataset and adapt to changes over time. Therefore, they are seen as a form of artificial intelligence. The main critique of ANNs is that they are a "black box" and researchers are therefore unable to determine which variable factors contain the most weight or to track how the network learns (Warwick 1995; Gahegan 2003). However, the applied focus here is on the determination of risk, not the method of training and the associated weights within the ANN. Therefore, it is suggested that ANNs be considered and examined due to their intrinsic ability to parse non-linear relationships, even if their ability to thoroughly test hypotheses is lacking.

Once models of risk have been developed it is necessary to test their validity. Many methods could be used for this. One example would be the incorporation of mortality from past events to examine whether they spatially match the increased areas of risk. Another method may be to employ emergency dispatch data for heat related conditions. Such dispatches seem to correspond to increases in thermal stress within the environment (Golden et al. 2008). Moreover, emergency dispatch data is intrinsically spatial. By employing known measures of human heat related morbidity/mortality it is possible to develop a fitness index of the models using kappa statistics. This would quantitatively demonstrate the validity and effectiveness of the model using a priori events.

The main product of using such analysis and blending it with current geospatial resources is to develop risk maps associated with disaster planning, such as an HRHD. The presentation of such maps is not straightforward, but their proper production is vital to the safety of the population under consideration. For instance, a vital question is "Which classification scheme to use when portraying risk in the city?". Should the map employ Jenk's optimization or rely on a quantiles approach? Should the maps be provided electronically on a GPS enabled mobile device to emergency response personnel? These are important questions and likely need to be fine tuned for the individual city based on the distribution of values associated with risk and the agency's resources.

10.6 Conclusions

There is an existing need to develop surveillance platforms for urban disaster preparation and mitigation. Extreme heat is currently the number one weather related killer and the impacts of it are likely to increase as a function of climate change (Meehl et al. 2000; Koffi and Koffi 2008; Kovats and Akhtar 2008). Therefore, a likely disaster for many cities, especially within temperate climates, is an HRHD. These particular events are underscored by their silence and the efficiency with which they exacerbate chronic and acute health conditions. It is vitally important to develop tools and incorporate existing technological assets to alleviate, and plan intervention during such occurrences.

It has been demonstrated through a conceptualization that geospatial technologies are viable systems with which to perform surveillance of HRHDs. Following the collection of data on the socioeconomic condition of vulnerable populations and incorporating it with thermal remote sensing resources it is likely possible to develop risk maps with dramatic specificity. The development of such maps can be performed using logistic regression, machine based learning tools, or other statistical methods of classification. However, any developed maps need to be tested against the realization of prior events so the fitness of the map can be quantified. It is hoped the conceptualization presented is of assistance to public health entities and researchers involved in disaster preparation and mitigation.

References

Aniello, C., Morgan, K., Busbey, A., & Newland, L. (1995). "Mapping micro-urban heat islands using Landsat TM and A GIS." *Computers & Geosciences* **21**(8): 965–969.

Basu, R., & Samet, J. (2002). "An exposure assessment study of ambient heat exposrue in an elderly population in Baltimore, Maryland." *Environmental Health Perspectives* **110**(12): 1219–1224.

Beniston, M. (2004). "The 2003 heat wave in Europe: A shape of things to come?" *Geophysical Research Letters* **31**.

CDC (1994). "Heat-related deaths-Philadelphia and United States, 1993–1994." *MMWR* **44**(25): 465–468.

Coll, C., V. Caselles, et al. (2007). "Temperature and emissivity separation from ASTER data for low spectral Contrast surfaces." *Remote Sensing of Environment* **110**(2): 162–175.

Cutter, S. L. (2003). "Social vulnerability to environmental hazards." *Social Science Quarterly* **84**: 242–261.

Ebi, K., T. Teisberg, et al. (2003). "Heat watch/warning systems save lives: Estimated costs and benefits for Philadelphia 1995–1998." *Epidemiology* **14**(5): S35–S35.

Ebi, K. L. and M. O'Neill (2008). "Climate change and human health: Risks and responses." *American Journal of Epidemiology* **167**(11): S135–S135.

Fouillet, A., G. Rey, et al. (2006). "Excess mortality related to the August 2003 heat wave in France." *International Archives of Occupational and Environmental Health* **80**(1): 16–24.

Fouillet, A., G. Rey, et al. (2008). "Has the impact of heat waves on mortality changed in France since the European heat wave of summer 2003? A study of the 2006 heat wave." *International Journal of Epidemiology* **37**(2): 309–317.

Furfey, P. (1927). A Note on Lefever's "Standand Deviational Ellipse". *American Journal of Sociology*. **94**(33), DOI: 10.1086/214336.

Gahegan, M. (2003). "Is inductive machine learning just another wild goose (or might it lay the golden egg)?" *International Journal of Geographical Information Science* **17**(1): 69–92.

Getis, A. and J. K. Ord (1992). "The analysis of spatial association by use of distance statistics." *Geographical Analysis* **24**(3): 189–206.

Glass, G. E., B. S. Schwartz, et al. (1995). "Environmental risk-factors for Lyme-disease identified with geographic information-systems." *American Journal of Public Health* **85**(7): 944–948.

Golden, J. S., D. Hartz, et al. (2008). "A biometeorology study of climate and heat-related morbidity in Phoenix from 2001 to 2006." *International Journal of Biometeorology* **52**(6): 471–480.

Harlan S.L., B., A.J., Prashad, L., Stefanov, W.L., & Larson, L. (2006). "Neighborhood microclimates and vulnerability to heat stress." *Social Science and Medicine* **63**: 2847–2863.

Jemal, A., M. Kulldorff, et al. (2002). "A geographic analysis of prostate cancer mortality in the United States, 1970–89." *International Journal of Cancer* **101**(2): 168–174.

Jin, M. L., J. M. Shepherd, et al. (2007). "Development of a parameterization for simulating the urban temperature hazard using satellite observations in climate model." *Natural Hazards* **43**(2): 257–271.

Jones, T. S., A. P. Liang, et al. (1982). "Morbidity and mortality associated with the July 1980 heat wave in St Louis and Kansas City, Mo." *JAMA* **247**(24): 3327–3331.

Kalkstein, A. J. and S. C. Sheridan (2007). "The social impacts of the heat-health watch/warning system in Phoenix, Arizona: assessing the perceived risk and response of the public." *International Journal of Biometeorology* **52**(1): 43–55.

Kalkstein, L. S. (1991). "A new approach to evaluate the impact of climate on human mortality." *Environmental Health Perspectives* **96**: 145–150.

Kalkstein, L. S., P. F. Jamason, et al. (1996). "The Philadelphia hot weather-health watch warning system: Development and application, summer 1995." *Bulletin of the American Meteorological Society* **77**(7): 1519–1528.

Katz, R. W. and B. G. Brown (1994). "Sensitivity of extreme events to climate-change – the case of autocorrelated time-series." *Environmetrics* **5**(4): 451–462.

Kimes, D., A. Ullah, et al. (2004). "Relationships between pediatric asthma and socioeconomic/urban variables in Baltimore, Maryland." *Health & Place* **10**(2): 141–152.

Klinenberg, E. (2001). "Dying alone: The social production of urban isolation." *Ethnography* **2**(4): 501–531.

Klinenberg, E. (2002). *HeatWave : A Social Autopsy of Disaster in Chicago*. Chicago, University of Chicago Press.

Koffi, B. and E. Koffi (2008). "Heat waves across Europe by the end of the 21st century: Multiregional climate simulations." *Climate Research* **36**(2): 153–168.

Kovats, R. S. and K. L. Ebi (2006). "Heatwaves and public health in Europe." *European Journal of Public Health* **16**(6): 592–599.

Kovats, S. and R. Akhtar (2008). "Climate, climate change and human health in Asian cities." *Environment and Urbanization* **20**(1): 165–175.

Kulldorff, M. (2001). "Prospective time periodic geographical disease surveillance using a scan statistic." *Journal of the Royal Statistical Society Series a-Statistics in Society* **164**: 61–72.

Kulldorff, M., T. Tango, et al. (2003). "Power comparisons for disease clustering tests." *Computational Statistics & Data Analysis* **42**(4): 665–684.

Liu, Y. B., Y. Yamaguchi, et al. (2007). "Reducing the discrepancy between ASTER and MODIS land surface temperature products." *Sensors* **7**(12): 3043–3057.

Lo, C. P. and D. A. Quattrochi (2003). "Land-use and land-cover change, urban heat island phenomenon, and health implications: A remote sensing approach." *Photogrammetric Engineering and Remote Sensing* **69**(9): 1053–1063.

Lorentzen, P., J. McMillan, et al. (2008). "Death and development." *Journal of Economic Growth* **13**(2): 81–124.

Mastrangelo, G., S. Hajat, et al. (2006). "Contrasting patterns of hospital admissions and mortality during heat waves: Are deaths from circulatory disease a real excess or an artifact?" *Medical Hypotheses* **66**(5): 1025–1028.

McGeehin, M. A. and M. Mirabelli (2001). "The potential impacts of climate variability and change on temperature-related morbidity and mortality in the United States." *Environmental Health Perspectives* **109**: 185–189.

Meehl, G. A., F. Zwiers, et al. (2000). "Trends in extreme weather and climate events: Issues related to modeling extremes in projections of future climate change." *Bulletin of the American Meteorological Society* **81**(3): 427–436.

Michie, D., D. J. Spiegelhalter, et al. (1994). *Machine Learning, Neural and Statistical Classification*. New York; London, Ellis Horwood.

Mirchandani, H. G., G. McDonald, et al. (1996). "Heat-related deaths in Philadelphia – 1993." *American Journal of Forensic Medicine and Pathology* **17**(2): 106–108.

Ned Levine, K. E. K. a. L. H. N. (1995). "Spatial analysis of Honolulu motor vehicle crashes: I. Spatial patterns." *Accident Analysis and Prevention* **27**(5): 663–674.

Newbern, C. (2008). Epidemiologist. D. Johnson. Philadelphia, PA.

NOAA. 2008. NWS Publications. http://www.nws.noaa.gov/

Ord, J. K. and A. Getis (1995). "Local spatial autocorrelation statistics – distributional issues and an application." *Geographical Analysis* **27**(4): 286–306.

Buechley, R. W., J. Van Bruggen, & L. E. Trippi (1972). "Heat island = death island?" *Environmental Research* **5**: 85–92.

Remais, J., S. Liang, et al. (2008). "Coupling hydrologic and infectious disease models to explain regional differences in schistosomiasis transmission in southwestern China." *Environmental Science & Technology* **42**(7): 2643–2649.

Robinson, P. J. (2001). "On the definition of a heat wave." *Journal of Applied Meteorology* **40**(4): 762–775.

Rosenzweig, C., Solecki, W.D., Parshall, L., Chopping, M., Pope, G., & Goldberg, R. (2005). "Characterizing the urban heat island in current and future climates in New Jersey." *Environmental Hazards* **6**: 51–62.

Saaroni, H., Ben-Dor, E., Bitan, A., & Potchter, O. (2000). "Spatial distribution and microscale characteristics of the urban heat island in Tel-Aviv, Israel." *Landscape and Urban Planning* **48**: 1–18.

Schaeffer, B., B. Mondet, et al. (2008). "Using a climate-dependent model to predict mosquito abundance: Application to Aedes (Stegomyia) africanus and Aedes (Diceromyia) furcifer (Diptera: Culicidae)." *Infection Genetics and Evolution* **8**(4): 422–432.

Semenza, J. C., McCullough, J.E., Flanders, W.D., McGeehin, M., & Lumpkin, J.R. (1999). "Excess hospital admissions during the July 1995 heat wave in Chicago." *American Journal of Preventative Medicine* **16**(4): 269–277.

Semenza, J. C., D. J. Wilson, et al. (2008). "Public perception and behavior change in relationship to hot weather and air pollution." *Environmental Research* **107**(3): 401–411.

Sheridan, S. C. (2007). "A survey of public perception and response to heat warnings across four North American cities: an evaluation of municipal effectiveness." *International Journal of Biometeorology* **52**(1): 3–15.

Sheridan, S. C., and Kalkstein, L.S. (2004). "Progress in heat watch-warning system technology." *American Meteorological Society BAMS* **85**: 1931–1941.

Smoyer, K. E. (1998a). "A comparative analysis of heat waves and associated mortality in St. Louis, Missouri – 1980 and 1995." *International Journal of Biometeorology* **42**(1): 44–50.

Smoyer, K. E. (1998b). "Putting risk in its place: Methodological considerations for investigating extreme event health risk." *Social Science & Medicine* **47**(11): 1809–1824.

Smoyer, K. E., Rainham, D.G., & Hewko, J.N. (2000). "Heat-stress-related mortality in five cities in Southern Ontario." *International Journal of Meteorology* **44**: 190–197.

Tiangco, M., A. M. F. Lagmay, et al. (2008). "ASTER-based study of the night-time urban heat island effect in Metro Manila." *International Journal of Remote Sensing* **29**(10): 2799–2818.

Unwin, A. (1996). "Geary's contiguity ratio." *Economic and Social Review* **27**(2): 145–159.

Vezzani, D. and A. E. Carbajo (2008). "Aedes aegypti, Aedes albopictus, and dengue in Argentina: current knowledge and future directions." *Memorias do Instituto Oswaldo Cruz* **103**(1): 66–74.

von Storch, H. and N. Stehr (2006). "Anthropogenic climate change: A reason for concern since the 18th century and earlier." *Geografiska Annaler: Series A, Physical Geography* **88A**(2): 107–113.

Voogt, J. A., & Oke, T.R. (2003). "Thermal remote sensing of urban climates." *Remote Sensing of Environment* **86**: 370–384.

Wang, K. C., J. K. Wang, et al. (2007). "Influences of urbanization on surface characteristics as derived from the Moderate-Resolution Imaging Spectroradiometer: A case study for the Beijing metropolitan area." *Journal of Geophysical Research-Atmospheres* **112**: D22S06, doi: 10.1029/2006JD007997.

Wang, W. W., L. Z. Zhu, et al. (2004). "An analysis on spatial variation of urban human thermal comfort in Hangzhou, China." *Journal of Environmental Sciences-China* **16**(2): 332–338.

Warwick, K. (1995). "A critique of neural networks for discrete-time linear-control." *International Journal of Control* **61**(6): 1253–1264.

Wong, D. W. S. (1999). "Geostatistics as measures of spatial segregation." *Urban Geography* **20**(7): 635–647.

Wong, D. W. S. (2003). "Implementing spatial segregation measures in GIS." *Computers, Environment, and Urban Systems* **27**: 53–70.

Wong, D. W. S. (2005). "Formulating a general spatial segregation measure." *The Professional Geographer* **57**(2): 285–294.

Zhang, T. L. and G. Lin (2008). "Identification of local clusters for count data: a model-based Moran's I test." *Journal of Applied Statistics* **35**(3): 293–306.

Chapter 11
Spatial Analysis, Policy, Planning, and Alternative Energy Production

James Pace and Jay D. Gatrell

Abstract This chapter examines the geography and capacity of solar and wind energy production. The chapter considers the efficacy of alternative energy policy at the state scale. The chapter determines that policy initiatives and incentives vary by state and that renewable portfolio standards are effective.

Keywords Energy policy · Renewable energy · Solar · Wind · GIS

Currently, the United States is the world's largest producer of electricity generating over 20% of electricity worldwide (Energy Information Administration (Energy Information Administration (EIA) 2006b). In 2005, the United States produced a total of 4,055 billion kilowatt hours and this production is expected to continue to rise at 2% per year on average as it has done in the past (EIA 2006b). Currently, the increase in production has kept up with demand (EIA 2006b). Since the energy crises of the 1970s, the United States federal government and most state governments within the United States have developed energy policies and incentives to spur development of alternative energy programs (Norberg-Bohm 2000). In addition, current energy sources have steadily become more socially, economically, and environmentally expensive. Therefore, alternative energy sources are going to prove to be increasingly important for both the future of national security, commerce and consumer electricity production. More importantly, the development of policies and planning practice surrounding alternative energy production will need to embrace a full complex of planning tools—including GIS and spatial statistics—to understand and explain the scale, scope, and efficacy of policy regimes.

The distribution of energy resources and their production currently is, and always has been, an important area of study within geography. Natural resources are found in various regions throughout the world and such resources are often distributed to distant populations far from the originating area. The central theme behind most energy studies is the spatial relationship between the resources and

---

J.D. Gatrell (✉)
School of Graduate Studies and Department of Geography, Geology, & Anthropology, Terre Haute, IN, USA
e-mail:jgatrell@isugw.indstate.edu/

J.D. Gatrell, R.R. Jensen (eds.), *Planning and Socioeconomic Applications*, Geotechnologies and the Environment 1, DOI 10.1007/978-1-4020-9642-6_11,

the logistics involved in bringing them to market. Currently, the majority of energy studies focus on the production of the most widely used energy sources such as coal, oil, and natural gas (Calzonetti 1981; Lydolph and Shabad 1960; Chapman 1974; Lins 1979; Heiman and Solomon 2004; Clements 1977; Sheskin 1978; Hudson and Sadler 1990; Elmes 1985, 1996). They have looked at (1) the general geography of the coal and natural gas industries as it pertains to discovery and the distribution of production, (2) the spatial constraints on delivery of the resources to market, (3) the location of multinational corporations, their facilities, both functionally and geographically, and (4) the overall pattern of land use and land cover development associated with energy generation (Calzonetti 1981; Chapman 1974; Lins 1979; Hudson and Sadler 1990; Lydolph and Shabad 1960; Clements 1977; Sheskin 1978).

This chapter examines the geographies of energy policy and production. An analysis of energy policies, incentives and production in the United States is needed in order to gain an understanding of how these policies affect incentives and address energy needs in order to assess the growth and use of sustainable energy practices. More importantly, this chapter demonstrates that alternative energy planning can benefit from spatial analysis and basic visualization techniques.

11.1 Background

Renewable energy resources have always been key to the survival of past societies. The use of water and wind to power mills and wood for fire are the most basic forms. The greatest difficulty in this regard has always been getting easy and plentiful access to our energy resources. Fossil fuels to date have been plentiful and are relatively easy to access, even as an imported good, but they are not renewable in a reasonable timeframe. As the political, economic and environmental circumstances in the world change, energy is now being sought that can be quickly renewable, environmentally friendly, and locally available. Future energy resources must be renewable without requiring human intervention and they must have a relatively short recovery period. Hydrological power almost meets these requirements but is limited by its proximity to existing rivers, its affect on natural habitat, and its dependency on continued rainfall and snowfall. Since most energy resources such as wood, coal and petroleum are depleted faster than they can be renewed only wind and solar power currently offer us a sufficiently renewable, locally available and environmentally friendly resource.

The history of wind energy generation had its beginnings in Persia and China around 2500 years ago and was used primarily for grinding grain and pumping water. The technology slowly developed, became more efficient, and entered into Europe during the 1000s. Wind power continued to pump water and grind grain and Holland began to use it to drain lakes and marshes for farm land (Nersesian 2006). However, the use of wind power was eventually supplanted by more reliable steam engines and fossil fuels and remained largely undeveloped until recently

(DOE 2005). Currently, wind energy is one of the fastest-growing renewable energy sources in the United States and the world (Menz 2005).

Solar energy generation had its beginnings in 1839 when French physicist Edmond Becquerel first discovered photovoltaics. He found that an electrical current could be increased in certain materials when exposed to light (Nersesian 2006). On a similar note, in 1861, Auguste Mouchout developed a steam engine powered entirely by the sun and in 1876, William Grylls Adams discovered that when selenium was exposed to light it discarded electrons, thus producing electricity. However, the development of reliable fossil fuels meant these ideas would not develop into a viable fuel source. It wasn't until 1921 when Albert Einstein was awarded the Nobel Prize in physics for his 1904 research on the photoelectric effect did a complete understanding of the generation of electricity through solar cells develop. Then in 1953, Bell Laboratories scientists Gerald Pearson, Daryl Chapin, and Calvin Fuller developed the first solar cell capable of producing an electric current. Solar cell efficiency gradually improved and solar technology was even used to power the Vanguard I satellite where it proved to be a major success (DOE 2006).

However, neither solar nor wind technology progressed substantially enough to gain widespread use by businesses and the average consumer. The energy crisis and oil embargos of the 1970s proved to be a major turning point. On October 17, 1973 the Organization of Arab Petroleum Exporting Countries (OAPEC) began limiting the export of oil to certain nations, the United States included, for various political reasons. The decrease in supply drove up prices and awakened the United States to its dependency on foreign oil (Szyliowicz and O'Neill 1975). This crisis sparked the search for alternative energy sources. This led to further research into renewable energy sources such as solar and wind power (Norberg-Bohm 2000). However, even with an understanding of this history, the United States and many other nations have continued to remain dependent on foreign controlled energy sources and non-renewable energy sources.

The commercialization and decentralization of the wind and solar energy industry started in the United States largely in part as a result of the Public Utility Regulatory Policies Act of 1978 (Norberg-Bohm 2000). This act required the state regulatory commissions to allow non utility companies to sell electricity from renewable energy resources back to the utility at an "avoided cost" rate (Menz 2005; Norberg-Bohm 2000). California went one step further and required utilities to buy back wind energy at a premium which is credited for its growth in the wind and solar energy sectors. Both industries have since increased production but as yet have been unable to gain a substantial share of the electricity market (Norberg-Bohm 2000). However, there has been a steady increase in the use and availability of incentives and policy for wind and solar energy development (DSIRE 2007).

In this chapter we will examine two key factors that affect wind and solar energy production—-geography and policy. Geography is vital to the success of these two technologies since they are dependant upon the location and concentration of wind and solar radiation. Public policy is also crucial in establishing current and future energy standards to best meet the public need. In the absence of such policies, efforts to promote alternative and renewable energy sources are easily marginalized in favor of the status quo. In addition to implementing policy, financial incentives are also necessary since both wind and solar energy production costs are currently slightly higher that those of conventional resources with the majority of costs incurred up-front. Without incentives, consumers may not be willing to make the financial investment toward renewable technologies when current technology is both cheaper and more accessible. The combined effort of the above measures will lead the development of alternative energy technologies to replace current ones before our current sources run out or become too expensive.

11.2 Physical Geography

The use of wind and solar energy are geographically more suited to certain areas of the United States than others. Naturally, an area that experiences more wind than another will make investing in wind energy production more economically feasible at that location. The same can be said of solar energy production. Within the United States the areas of greatest wind production potential are in the Great Plains and upper Midwest, along coastlines that experience trade winds or westerly winds, mountain tops, mountain passes or interior valleys that develop thermal low pressure areas (Simon 2007; Menz 2005; Christopherson 2006).

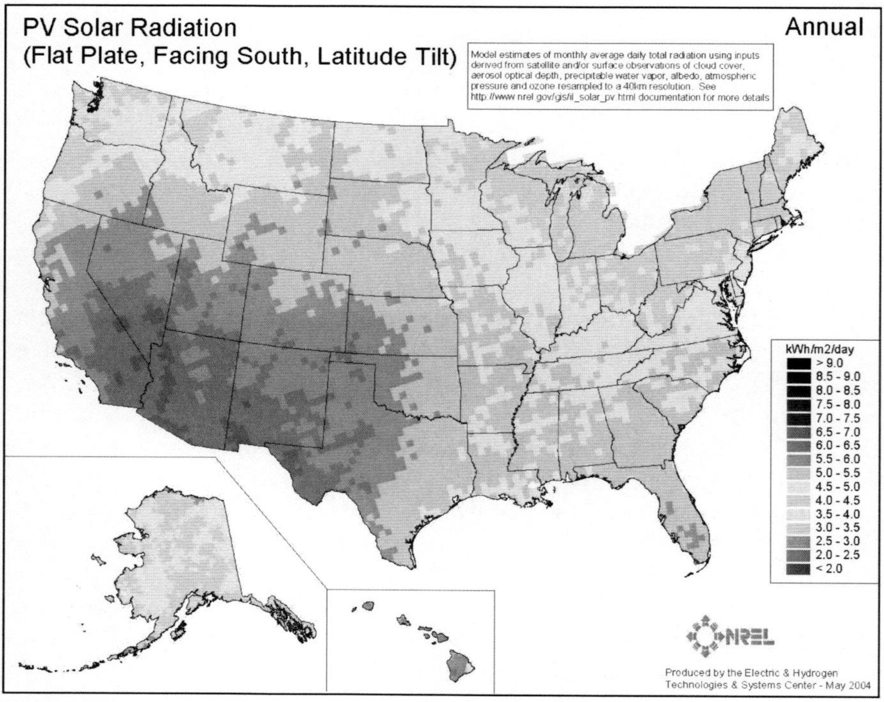

Fig. 11.1 NREL produced estimate of photovoltaic solar radiation

Areas of greatest solar energy production are those lying closest to the equator where the concentration of solar radiation is the greatest (Taylor 2006). In addition to latitude, consideration must be given to topography, atmospheric conditions, site orientation and shading from structures or trees (Dogniaux 1994). These factors; topography, climate, weather and cloud cover all greatly affect a locations relative productivity (Taylor 2006). In the United States, the southwest has the greatest potential with net solar radiation values decreasing as it radiates from this region (NREL 2007a). However, enough solar energy falls at every location within the United States to make solar power feasible (Christopherson 2006). Therefore, there is the potential to meet current electrical needs with wind and solar energy. It only remains to be seen if the cost of production and transmission can be brought down to match those of fossil fuel sources.

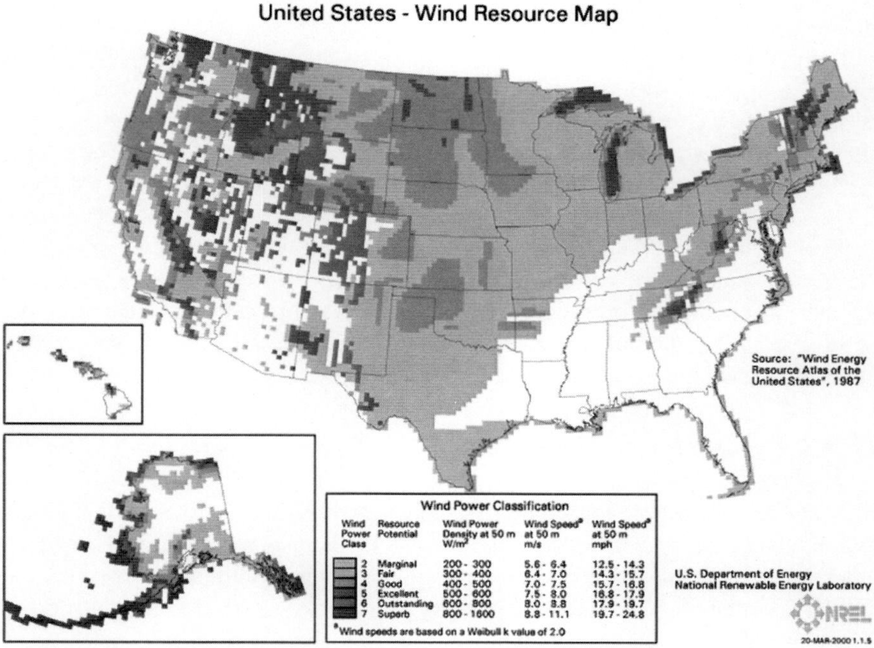

Fig. 11.2 NREL produced wind resource map

11.3 Public Policy: Crises, R&D, and Policy Initiatives

In addition to the physical geographies that provide the opportunity for effective and efficient wind and solar production, this study emphasizes that public policy and financial incentives are a determinant of alternative and renewable energy production in the United States. Specifically, this research will consider the historical role policy initiatives have played in the development of industry (particularly in response to crises), commercialization of core technologies (Research and Development), and current regulations and incentives.

11.3.1 Crises

The formation of energy policy in the United States can be attributed to a series of energy crises. Until the early 1970s, energy policy and the search for alternate forms of energy production were of little importance to the United States federal government (Bezdek 2007). Funding for energy projects was spread across multiple federal agencies: the Atomic Energy Commission, (AEC) the Department of the Interior, the Department of the Treasury, and the State Department, all with varying interests and goals (Bezdek 2007). The first catalysts for change began with the Arab Oil Embargo of 1973–1974 and the Iranian Revolution of 1979 (Bezdek 2007).

Both of these events caused the disruption of oil supplies and resulted in the rapid rise of prices for oil and other fossil fuels (Gan 2007). In the 1990s international treaties such as the Kyoto Protocol (1997) helped to push the development of alternative fuels with the goal of reducing global warming (Gan 2007). The price of oil then spiked again in 1992 due to the events of the Persian Gulf War (Norberg-Bohm 2000). In 2000–2001 a different and localized energy crisis took place in California due to problems of deregulation and business corruption. Most recently, the United States has experienced steadily increasing prices for oil due in part to the current conflict in Iraq. Every energy crisis has brought about the increased awareness and development in energy policy, incentives, research and development in renewable energy, all with varying degrees of utility.

11.3.2 Research and Development

Bezdek (2007) observed that, since the energy crisis of the 1970s important federal research and development support has usually favored solar and wind energy technologies rather than nuclear and fossil fuel technologies. But the major recipient of federal subsidies has been the oil industry which takes in over half of all federal subsidies. In addition to the fact that new energy technologies must compete based on market prices, there are some additional features that solar and wind will need to address in addition to low price. These unaccounted costs include: security of fuel supply, fuel flexibility, distribution and modularity, and environmental impact (Norberg-Bohm 2000). Therefore, renewable energy will compete not only on the cost of energy, but on the value provided to the customer and the benefits to the government.

First and foremost, the growth of wind and solar energy in the United States and the world is robust. From 1990 to 2003 wind generation capacity grew from 1525 megawatts (MW) to 6374 MW, an increase of 282%. In percentage terms, wind generating capacity is growing faster than even conventional power sources (Menz 2006, EIA 2005). In 2006, the National Renewable Energy Laboratory

(NREL) estimated that current installed wind production capacity was had grown enough to power 2.3 million households (NREL 2007b). Photovoltaic cell and module domestic shipments reached a record high of 134,465 peak kilowatts in 2005, a 72% increase from the 2004 milestone 78,346 peak kilowatts, and that was an increase of more than 176% from the 2003 level (EIA 2005).

Even with such robust growth the solar and wind industries are still a minor contributor of energy in percentage terms accounting for about 2% of electricity needs per year. The first major obstacle is the relative low cost and abundance of conventional fuel such as coal. The literature on solar and wind energy generation for the most part agrees that the inability of renewable energy to break into the electricity market is caused by the higher cost of renewable energy compared to conventional energy (McVeigh et al. 2000; Darmstadter 2000; Taylor and Van-Doren 2002; Nogee et al. 1999). Coal has and most likely will remain the main source of electricity generation in the United States and has accounted for more than 50 % of electricity consumption since 1980 (EIA 2004; Menz 2005). Current coal reserves can provide at least 400 more years of electricity at our current consumption rate and with continuing technological advances in coal combustion and processing technologies. Due to this fact, the United States Department of Energy forecasts a continual drop in the price of coal for the foreseeable future and expects it to still account for nearly 50% of net electric generation in 2025 (EIA 2003; Menz 2005).

Therefore, even with policies and incentives to stimulate growth, the overall share of electricity produced from these two sources has not changed (Menz 2005). This can be attributed to the high capital and transmission costs for renewable technologies and the relatively moderate costs of electricity produced from fossil fuels (Gan 2007). In most energy forecasts, conventional energy production costs were assumed to have risen much larger allowing both solar and wind technologies to be competitive in the market. However, this has not been the case. In addition to the unexpected decline in conventional energy prices, government subsidies have been biased toward conventional sources, thus increasing the difficulty for renewable energy sources to compete in the electricity market.

As an example, 12 billion was spent on renewable energy research and development between 1973 and 1998, while the federal government spent roughly 22 billion on nuclear energy and 21 billion on fossil fuels (Bezdek 2007). However modest the growth may be in terms of total percentage of electricity production, the current levels of alternative energy production are rising quickly. In any case, as long as wind and solar energy are more expensive to produce, public energy policy and financial incentives will be the major short-term force behind production levels and support for these technologies.

11.3.3 Regulations

The implementation of regulations and policies can be an essential component to the development or deployment of any alternative technology that cannot initially compete on the open market. In this light, both wind and solar technologies are aided by these measures to fuel their growth. The types of regulations enacted are varied. Many regulations are passive and promote new technology through construction and design policies, contractor licensing, equipment certifications, and energy generation disclosure rules. The aim of these regulations is to provide information to the public on the source of energy and the quality of the labor or equipment (DSIRE 2007).

There are also a set of regulations that actively seek to introduce new technologies by adjusting market forces. These are: aggregation policies, green power purchasing, net metering, and renewable portfolio standards. Aggregation policies require states, businesses, or cities to purchase a predetermined amount or percentage of energy from renewable resources effectively making a certain region a "green community". In some states, utilities are required to offer a green power option to its customers. This allows customers to purchase power through the utility from green or renewable resources bringing demand for the technology from the state and local level down to the individual consumer. Net metering rules require utilities to purchase energy from individuals that generate their own power from wind or solar technologies. This allows and encourages individuals to generate as much energy as possible but also allows them to be connected to the grid and buy electricity from the utility when their systems are not meeting their energy demands. Renewable Portfolio Standards (RPS) go an additional step further by requiring a utility to generate a certain percentage of its power from renewable technologies each year or be fined. This method creates an instant demand for renewable energy (DSIRE 2007).

A number of studies have been conducted that evaluate the policies and regulations put in place to support wind and solar technologies. Lancaster and Berndt (1984) and Sawyer (1983) are some of the first researchers on record attempting to measure and quantify the progress of the first enacted policies and incentives designed to spur growth in the industry. They focused on household incomes, state and federal income taxes, sales tax, property tax and the cost of electricity.

Bird et al. (2005) studied key factors in 12 states that had a significant amount of developed or planned wind energy. They studied RPSs, federal and state financial incentives, consumer demand, natural gas prices, and wholesale market rules. Lie (Effect of States Incentives on Renewable Energy Generation, unpublished paper, Georgetown University, 2003) conducted quantitative analyses of the effects of state renewable energy policies. Langniss and Wiser (2003) performed a case study looking at the implementation of Renewable Portfolio Standards (RPS) in Texas. Recently, more case studies have been conducted that compare the United States green energy policies with those of other countries. Agnolucci (2006) evaluates the success of Germany and its renewable energy policy. Gan et al. (2007) compared Germany, Sweden, the Netherlands and the United States in order to look at a broad array of policies in different governing systems. McVeigh et al. (2000) compared actual performance of the technologies against projections provided to the public to construct policy.

11.3.4 Incentives

The last tool used to promote renewable energy production is to provide financial incentives. These incentives fall into two basic categories that are offered at the individual and corporate level: incentives for production and incentives for consumption. Production incentives typically provide a rebate or repayment based upon the amount of electricity generated per kilowatt hour. Consumption incentives include tax breaks and special grants or loans to be used to purchase the systems.

Gouchoe et al. (2002) examined 10 incentive programs in 6 states in an attempt to explore their effectiveness at stimulating the use of renewable energy technology. Rich and Roessner (2000) focused on effectiveness of federal tax credits for the commercialization of the solar industry. Goldberg (2000) looked at subsidies, tax incentives and insurance between 1943 and 1999 and found that the nuclear energy industry received 145 billion dollars compared to 6 billion dollars for solar and wind energy. Lewis (2007) examined the context that financial incentives are applied in conjunction with certain policies. Bezdek (2007) most recently, conducted research that looked at the past 50 years of incentives and policy.

The majority of these studies all indicate that any hindered growth of solar and wind is the lack of consistency and stability in policy and incentives (Norberg-Bohm 2000). A large part of the success of the German wind and solar programs come from the long-term contracts and policies that were adopted. Stop and start policies and incentives simply add to the uncertainty of the programs and ultimately result in higher costs, in part because investors and producers are dependant on financial and social indicators about the continued support and long-term commitment of the government toward the development of these resources (Lin 2007). This cycle is evident is a study by Lewis (2007) where he compared the United States and Indian markets to those of more stable German, Spanish and Norwegian markets where these technologies have flourished.

Deyette et al. (2003) assess the states support of renewable energy sources by comparing commitments in production, funding for renewable electricity, each state's renewable electricity generation and potential. Lewis (2007) looks at the motivations behind different countries in developing wind energy systems and the steps taken to grow the manufacturing industry.

In the United States there are two methods used to affect change on the state or national scale: (1) Legislation can be passed which will dictate the framework for all participants of energy production, and (2) financial incentives for energy production can be provided. Financial incentives are offered at the federal, state, utility and local level while policy is handled almost exclusively at the state and local levels. On the surface, incentives and policies appear to be similar across all states but upon closer examination there are substantial differences. The federal government, individual states and local municipalities all vary in their regulatory and financial circumstances and different variations of the same concept impact the efficacy of each policy or incentive toward promoting green electricity (Menz 2005). Tax policy has been, by far, the most widely used incentive mechanism, accounting for 281 billion (43.7%) of all Federal incentives (Bezdek 2007).

11.4 Methods

In order to examine the spatial dynamics of production and growth, we utilize the following variables: state and national energy incentive types (producer, consumer, both, and renewable portfolio standards), annual energy production and solar or wind capacity levels. Data sources include the Energy Information Agency of the United States Department of Energy, the Database of State Incentives for Renewables and Efficiency (DSIRE), and the American Wind Energy Association (AWEA). Incentive data for wind and solar technologies were obtained from the Database of State Incentives for Renewables and Efficiency (DSIRE). Many types of incentives exist for both technologies, all with different requirements, limitations, and possible effectiveness. Therefore, a state is considered to have had an incentive in effect for a given year as long as at least one incentive was offered, regardless of the type.

Wind energy data were available from the EIA in megawatts and shows how much energy can be generated by the collective turbines per second. To actually measure how much energy was generated you would multiply this capacity number by time. The resulting figure would be measured in mega-watt hours. Therefore, wind energy is measured in terms of capacity not the amount produced.

Solar energy capacity or generation figures are much more difficult to determine. Actual production information is not representative of the growth in the industry since production can vary widely depending on the location and the type of solar collector used. Also, solar energy is largely used by individual property owners rather than large corporate entities. Therefore, while wind farms are scattered across the United States, solar farms with measurable output are concentrated in only two or three states. This has likely occurred because large scale generation and transmission of solar energy is most profitable in only two or three states.

However, solar technology is widely used and one measure that the EIA has used to track this industry is the square footage of solar thermal panels shipped within the United States. The chapter argues that measuring shipments of solar thermal panels is a good proxy for solar energy capacity and production as it measures a technology that is both developed and one that there is current demand for. Therefore, we can determine how much capacity (square feet) is being added each year for each state by looking at how many square feet were shipped to each state. Some of these shipments may be to replace existing systems but all will be considered as growth since if they were not replaced it could be assumed that the capacity would have decreased and therefore would have affected capacity in a negative fashion.

National energy wide figures were reported in megawatt hours or the amount that was actually produced and not the capacity of the entire energy sector. These data will allow for a comparable contrast of growth in percentage terms. The final measure used in the analysis was whether or not a state had Renewable Portfolio Standards (RPS) in effect during 1999 and 2006. The timeframe of this study is also appropriate since it begins at a time when each technology had developed fully enough that incentives could realistically had the desired effect to promote growth within the industry.

11.4.1 Visualization and Analysis

The data will be manipulated to obtain both growth and location quotients. The obtained values will be mapped for visualization purposes and used as the basis for subsequent chi-square analysis.

Growth and location quotient will be used to determine the extent and concentration of growth of wind and solar energy generation (see Ó hUallacháin and Reid 1991). These methods of analysis are normally used to measure different aspects of regional employment growth but will also be suitable to measure energy growth. The growth quotient was determined by first calculating the normalized growth of solar and wind energy production at the state and national level and also the normalized growth of total energy produced at the state and national level. The state growth rate for wind and solar and national growth rate for wind and solar are then subtracted from the rate of growth of total energy production at the state and national levels respectively. The growth quotient of the wind or solar energy production is the ratio of these two normalized growth rates. A value of 1 signifies that a state has experienced growth in wind or solar consistent with the national average. Values below 1 indicates a below average growth rate and above 1 indicates an above average growth rate.

The location quotient was determined by dividing state level solar and wind capacity by national solar and wind capacity. Then total state energy production is divided by total national energy production. The location quotient is the ratio of these two values. This quotient allows for a comparison to be made of each state. It shows which states have the greatest concentration of wind and solar capacity compared to their own level of energy production and the national average. Comparing each of these measures will allow for the accurate measurement of growth on the state level and will then give an accurate comparison to both the national growth level and the growth level of each technology.

(1) Location quotient

$$\frac{(\mathrm{e}2006/\mathrm{E}2006)}{(\mathrm{n}2006/\mathrm{N}2006)}$$

(2) Growth quotient

$$\frac{\left(\frac{e2006 - e1999}{e1999}\right) - \left(\frac{E2006 - E1999}{E1999}\right)}{\left(\frac{n2006 - n1999}{n1999}\right) - \left(\frac{N2006 - N1999}{N1999}\right)}$$

where:

e = state solar/wind energy production
E = total state energy production
n = national solar/wind energy production
N = national energy production

Chi-square analysis required that location quotient, growth quotient, and potential data were transformed into nominal classes. To accomplish the data transformation, a standard natural breaks approach was used to reclassify the data. The growth quotient results were plotted on a line graph and natural breaks were assigned where the largest changes in slope were observed. The resulting groups that were created were: very high, high, medium, low and very low. The location quotient results were plotted on a line graph and breaks were assigned where the largest changes in slope were observed. The resulting groups that were created were: very high, high, medium, and low. A very low category for solar was not assigned since there were no observed values below zero.

Wind and solar resource potential were grouped into categories of low, medium and high. Solar energy resource potential for each state was divided into classes and measured in average watts per meter squared. Wind energy resource potential for each state was also divided into classes and measured in average power class per producing squared kilometer. These data had to be created using GIS since the geographic potential of wind and solar energy for each state was not available on a state scale. However, NREL provides resource potential maps based on the geographic availability of each resource. This information was put into ArcGIS and clipped to state boundaries. The average resource potential for each state was then calculated and the average watts per meter squared or average wind class per square meter was determined. Renewable portfolio standard (RPS) were grouped into two categories; states that had the incentives in effect during 2006 and those that did not.

Finally, incentives were divided into four categories: states that offer producer incentives, consumer incentives, both or none at all. A more detailed analysis of the content of each type of incentive and its effect on production is beyond the scope of this paper but applicable for future studies.

11.5 Results

11.5.1 Energy Resource

Figures 11.3 and 11.4 show the average resource availability for wind and solar energy production. Even though these data had been generalized to the state level, the same overall patterns associated with each resource (as compared to Figs. 11.1 and 11.2) are still present indicating that the representation is still accurate given the generalization. Based on the average wind power class most of the states west of the Rocky Mountains would be adequate for wind power generation. The AWEA states that any location with a wind resource class of three (3) and above is suitable for commercial wind energy production (AWEA 2008). The New England states, states bordering the Great Lakes, and Hawaii also have promising wind resource potential. This leaves the southeast and a section of the Midwest from Missouri to the east coast lacking in wind resource. However, this assessment is solely based upon average potential. In referencing Figs. 11.2 and 11.4, some states that have low

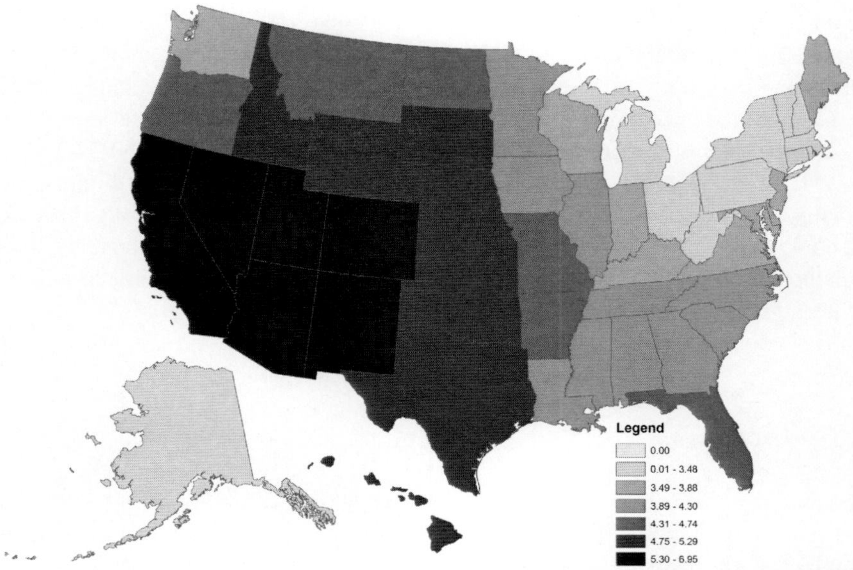

Source: National Renewable Energy Laboratory (NREL)

Fig. 11.3 Average solar insolation by state

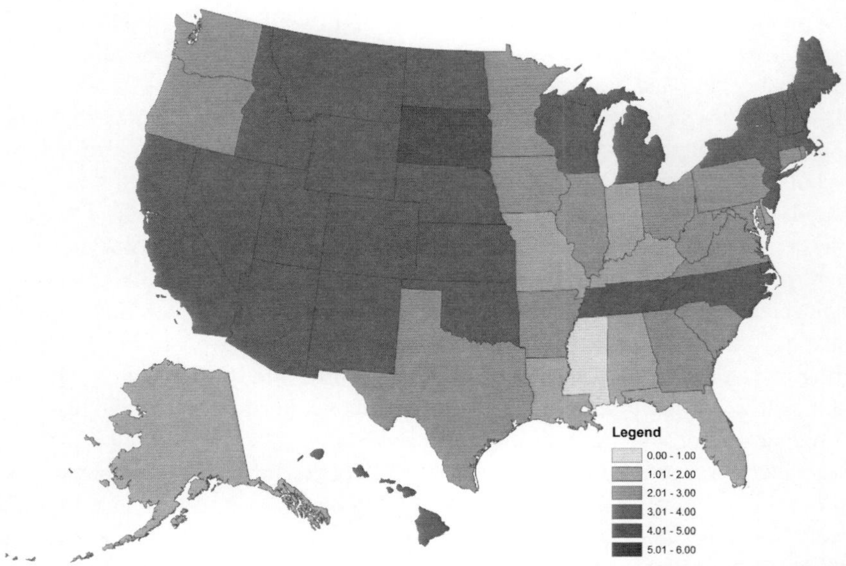

Source: National Renewable Energy Laboratory (NREL)

Fig. 11.4 Average wind power class by state

overall wind potential in Fig. 11.4 do have areas that are suitable for wind energy production when referencing Fig. 11.2. Since wind generation would be restricted to these specific areas, growth would also be limited and therefore the state average per meter for that resource is low.

The solar energy resource map also follows the same pattern as the NREL maps, Figs. 11.1 and 11.2, of highest resource potential in the southwest and diminishing from this area outward. Since solar radiation falls at every location the solar resource potential map is a better representation of the resource available on a state by state basis than wind is. Even though, both maps are a good representation and will be adequate to perform the analysis needed.

11.5.2 Incentives

The next maps Figs. 11.5 and 11.6 show the availability of incentives for wind and solar energy production. The overall availability of incentives for wind and solar during 1999 was relatively small. Gas prices averaged around a dollar and a half, the economy was booming as was the stock market, and it was also a pre-September 11 world and economy. Therefore, the overall lack of availability of incentives offered for wind and solar energy during 1999 is of little surprise.

Less than half of states offered consumer incentives for wind during 1999. Production incentives were even rarer, being only offered by one state, Minnesota. Many of these states naturally have areas of high wind resources. Therefore, it seems that where wind resources were adequate, incentives were generally offered to encourage continued or increased growth.

By 2006, most states had some form of consumer incentive in place with a handful of states offering production incentives also. A number of factors may have contributed to this expansion, the most probable explanation being the new political landscape that had been formed due to increased energy prices and a shift in public opinion to "go green" and focus on renewable energy technologies. Even with all of the effort made to "go green" only consumer incentives have been put into place across the board. The distribution of production incentives don't appear to have any pattern to them other than that the Pacific Northwest had many states offering incentives. In the southeast where there is little wind capacity, they adopted wind incentives. Therefore, it seems that other factors, in addition to wind resource potential, are the driving agents for wind incentives.

Solar incentives also followed the same overall pattern of those states that offered wind consumer incentives. Only the states of Minnesota, Mississippi and Virginia had solar incentives instead of wind incentives. This distribution is unexpected as it would be most probable to only see states with high solar resource potential with solar incentives. Since the states offering both wind and solar incentives in 1999 are so similar, the creation of these incentives are most likely the result of politics and not market based incentives or incentives created due to location of resources. No producer incentives were offered during 1999.

(A)

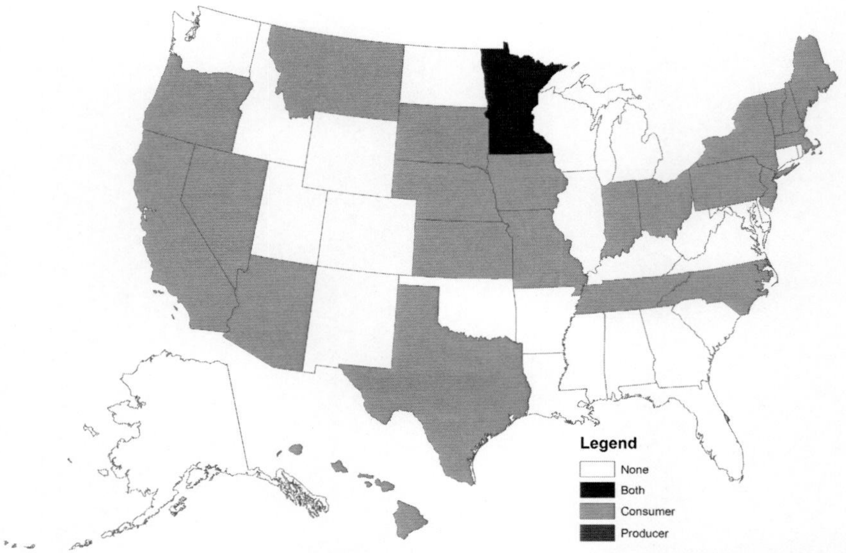

Source: Database of State Incentives for Renewable & Efficiency (DSIRE)

(B)

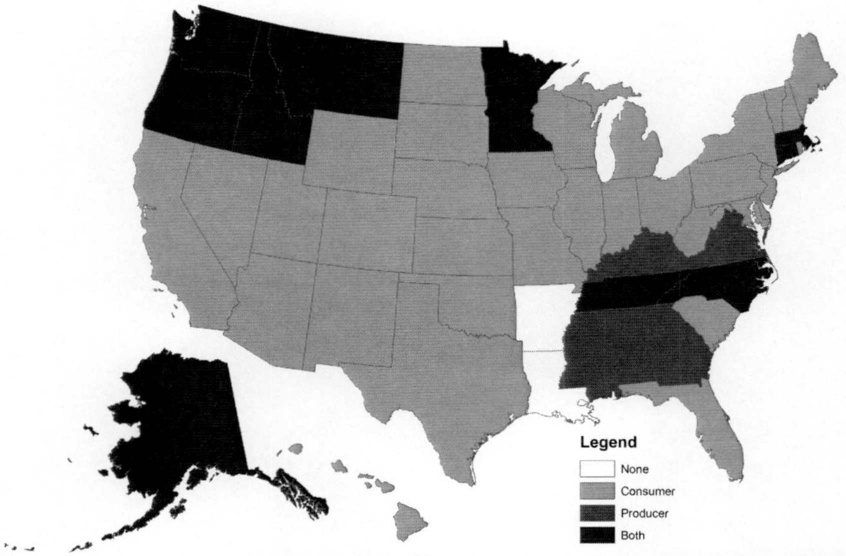

Source: Database of State Incentives for Renewable & Efficiency (DSIRE)

Fig. 11.5 Wind incentives by state, 1999 (**A**) and 2006 (**B**)

(A)

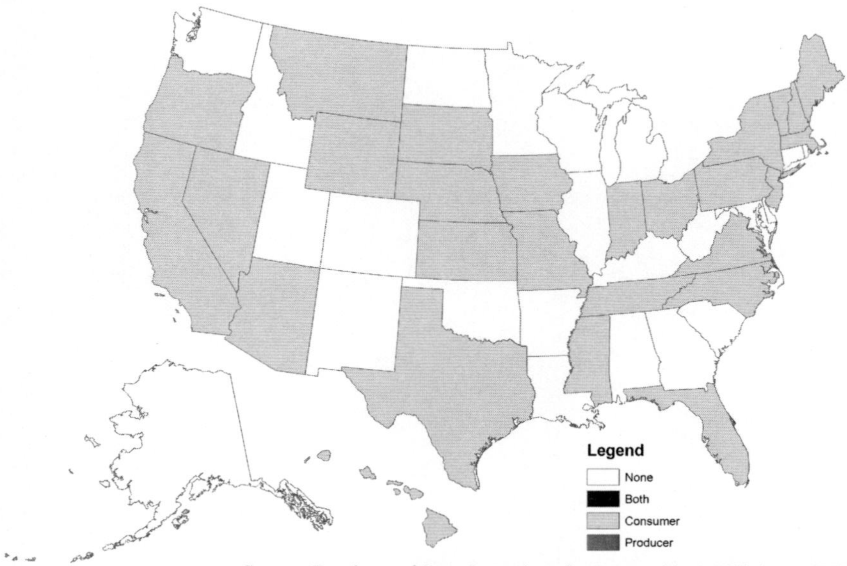

Source: Database of State Incentives for Renewable & Efficiency (DSIRE)

(B)

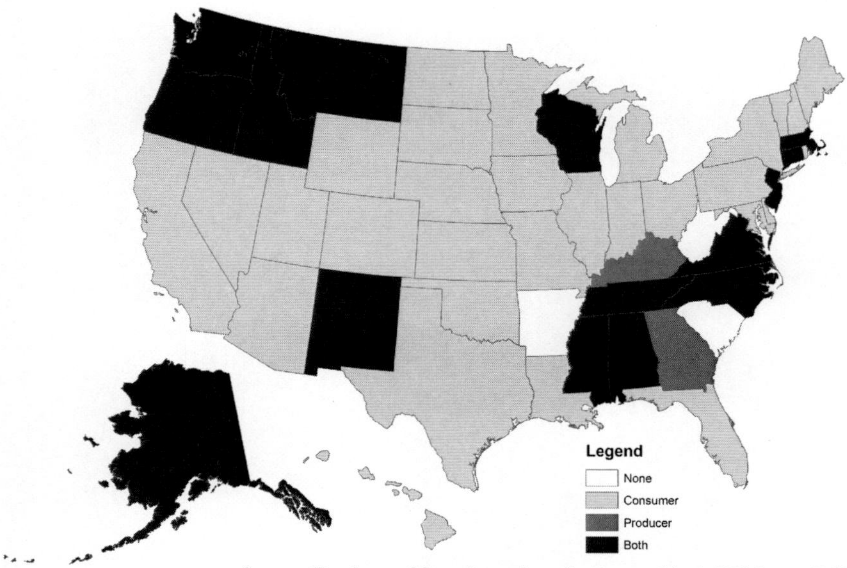

Source: Database of State Incentives for Renewable & Efficiency (DSIRE)

Fig. 11.6 Solar incentives by state, 1999 (**A**) and 2006 (**B**)

By 2006, solar consumer incentives were offered in almost every state, much the same as was seen with wind. Solar producer incentives saw the same comparatively modest increase that wind energy had, but with an increased presence in the southeast. Therefore, states have made a consistent effort to provide incentives to the consumers of wind and/or solar technology to help with the purchase of equipment. However, very little effort is being made to encourage actual production. Therefore, producer incentives for solar and also wind must either be viewed as unnecessary, not productive, too expensive, or are simply ignored as a viable option. However, there is no unambiguous explanation why consumer incentives are more heavily favored over production incentives. It is likely the product of many social, economic, and political factors that will vary from state to state and without an in-depth reading into the energy history of each state a plausible explanation is a loosely based educated guess.

11.5.3 Renewable Portfolio Standards

The distribution of states with renewable portfolio standards (RPS) also did not exhibit any meaningful patterns as Figs. 11.7 and 11.8 illustrate. The creation of renewable portfolio standards (RPS) is political and a result of economic and social pressures. While renewable portfolio standards (RPS) all have similar traits, states can draft very different versions of each and may not consider the wind or solar resource potential in the state but rather some other technology. A state, therefore,

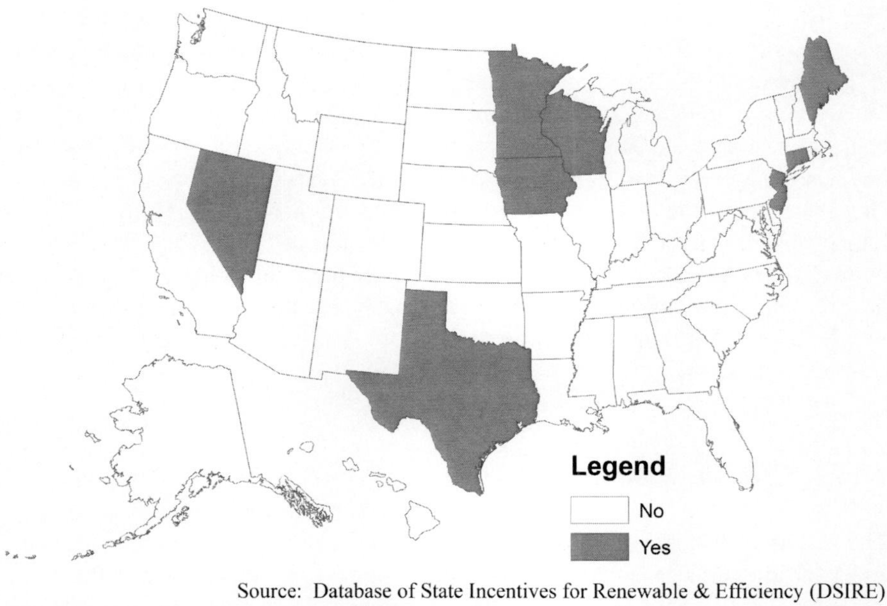

Source: Database of State Incentives for Renewable & Efficiency (DSIRE)

Fig. 11.7 Renewable portfolio standards by state, 1999

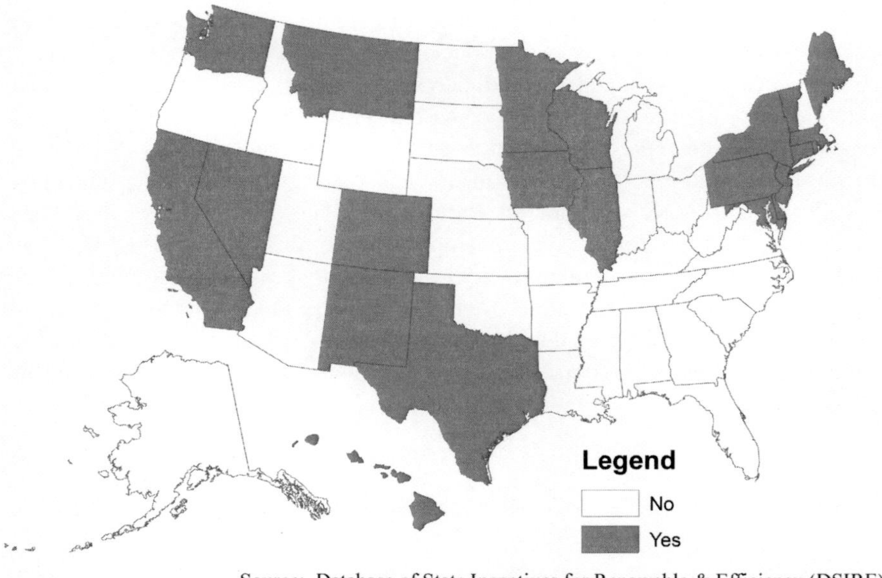

Fig. 11.8 Renewable portfolio standards by state, 2006

may create a RPS in order to promote only one or two technologies and solar and wind may not have been one of those technologies that the RPS would have benefited. However, based on the growth quotient, states with the lowest solar or wind resource potential are also the states experiencing the most growth in wind and solar.

In addition, RPS require a certain percentage of energy production to come from renewable technologies. They typically do require that this percentage of energy to come from within the state. Therefore, it is possible and to be expected that utilities would buy renewable power from across state lines to fulfill this requirement. However, due to the high cost of electricity transmission, such transmission over long distances is not expected. In 1999, few states had instituted RPS. By 2006, the number of states with RPS increased substantially but it did not show any apparent pattern. The divide is likely between the states that like the idea and/or want growth in energy production from renewables, or those that have no resources or do not want to regulate it.

11.5.4 Growth Quotient

The results of the growth quotient provided two very different outlooks on wind and solar growth. The results of the solar growth quotient, Fig. 11.9, illustrates that a couple of states with very high values that skew a correct interpretation. This is most likely the case because those states with the highest growth had little or no

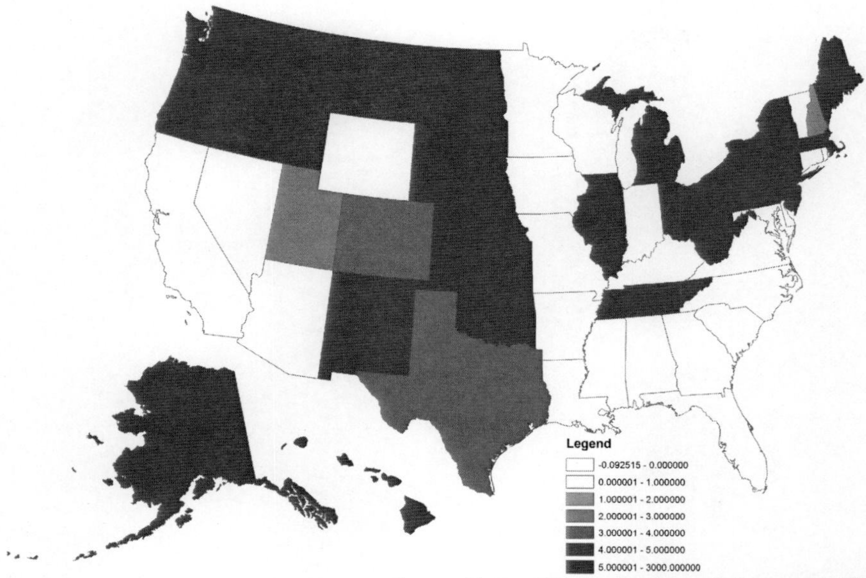

Source: National Renewable Energy Laboratory (NREL)

Fig. 11.9 Wind growth quotients

solar capacity in 1999. Therefore, with a decent increase by 2006, it appears that those states had enormous growth when, actually, they added only a modest amount of squared footage. However, most notable is that the states that are often viewed as having a high concentration of solar are not those states represented as having large growth. Rather, the opposite is true. It seems as though many of the "solar" states may have already experienced their greatest growth prior to the beginning of the study period and are now more matured and experiencing slowed growth rate.

Meanwhile, those states with the greatest solar growth were those for which solar technology was new or now efficient enough to be suitable for locations with solar resources that are less than exceptional. However, the growth quotient map for wind (Fig. 11.10) capacity mirrors that of wind resource capacity. States with the greatest growth were in the northwest, along the Rocky Mountains and eastward toward the Missouri River, along the Great Lakes and in the northeast. States such as California that already have high wind capacity experienced low growth as this market is likely matured and growth can be expected to be steady.

11.5.5 Location Quotient

The solar location quotient exhibits many of the qualities that one could expect in regard to wind and solar capacity distribution (see Figs. 11.11, 11.12, 11.13 and 11.14). In 1999, the concentration of solar panels was heavily in California, Nevada,

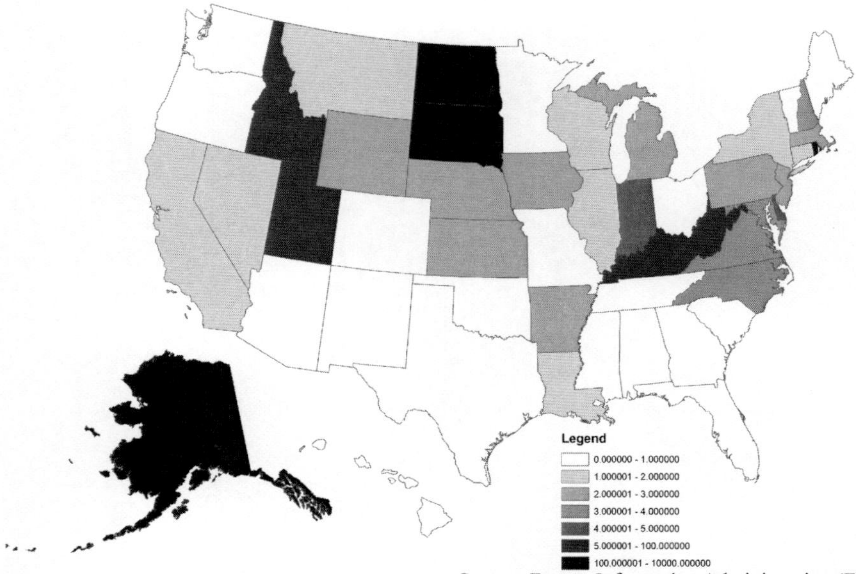

Source: Energy Information Administration (EIA)

Fig. 11.10 Solar growth quotients

Arizona, and Florida with much of the rest of the United States at fairly low levels. By 2006, this trend continued pushing the location quotient more heavily in favor of California, Nevada, Arizona and Florida. States such as Colorado, New Mexico, Illinois, Virginia, Connecticut, New York, and Delaware also showed overall increase just not as substantial an increase. This shows that while the states with higher than average solar resource didn't experience large growth during 1999–2006 they did have the largest holdings of solar panels. Therefore, any state that may have already gone through its initial growth phase for solar will not stand out in the growth quotient.

However, the location quotient will show how the overall capacity of solar is distributed against established energy production. (See Appendix B for the location quotient values for all 50 states) The wind location quotient in 1999 heavily favored California, Minnesota, Iowa, and Wyoming with much of the United States having relatively little to no capacity. By 2006, California had lost some of it prominence while almost all the states west of Kansas including the New England states saw substantial increases. This increase almost mirrors the wind resource map. However, the New England states and those around the Great Lakes do not show the same increased concentration. We know from the growth quotient that those areas did increase, just that they didn't increase substantially enough in terms of aggregated wind capacity. This could be due to higher concentration of people and fewer places to locate wind farms as problems of a not-in-my-back yard (NIMBY) mentality are likely to be the culprit.

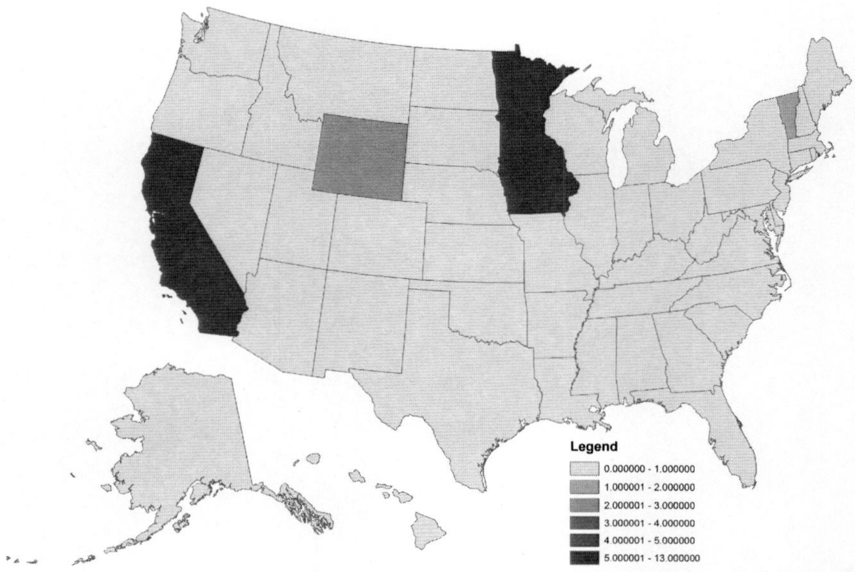

Source: National Renewable Energy Laboratory (NREL)

Fig. 11.11 Wind location quotient, 1999

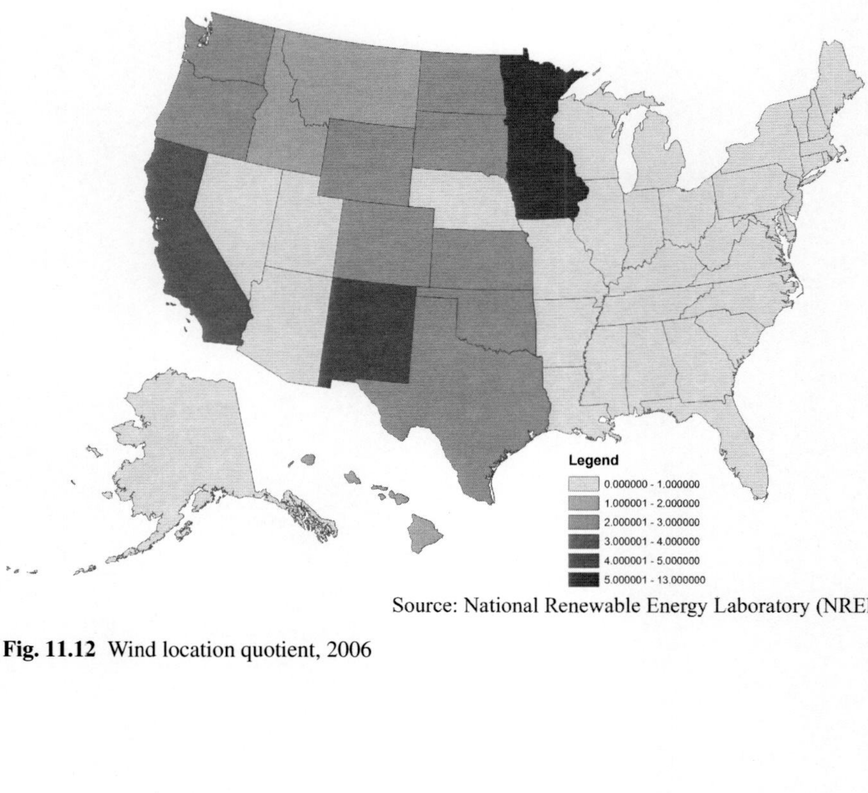

Source: National Renewable Energy Laboratory (NREL)

Fig. 11.12 Wind location quotient, 2006

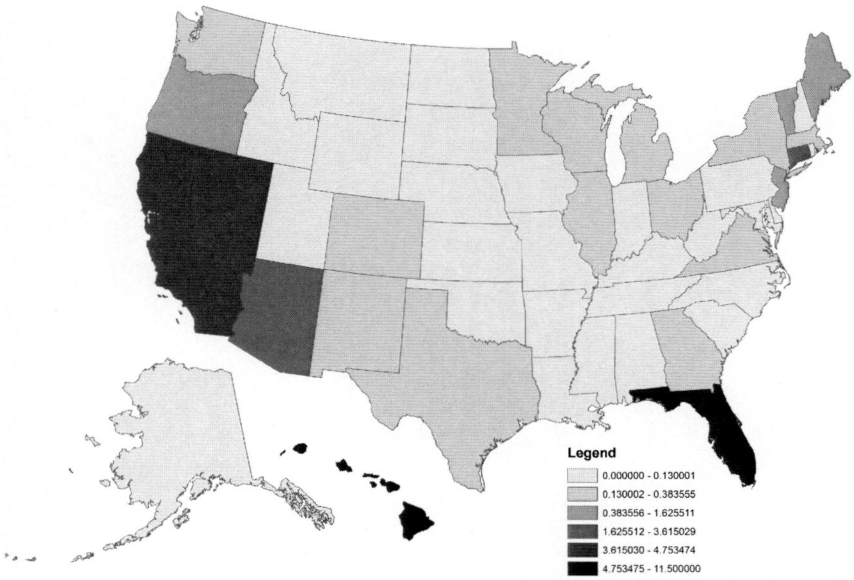

Source: Energy Information Agency (EIA)

Fig. 11.13 Solar location quotient, 1999

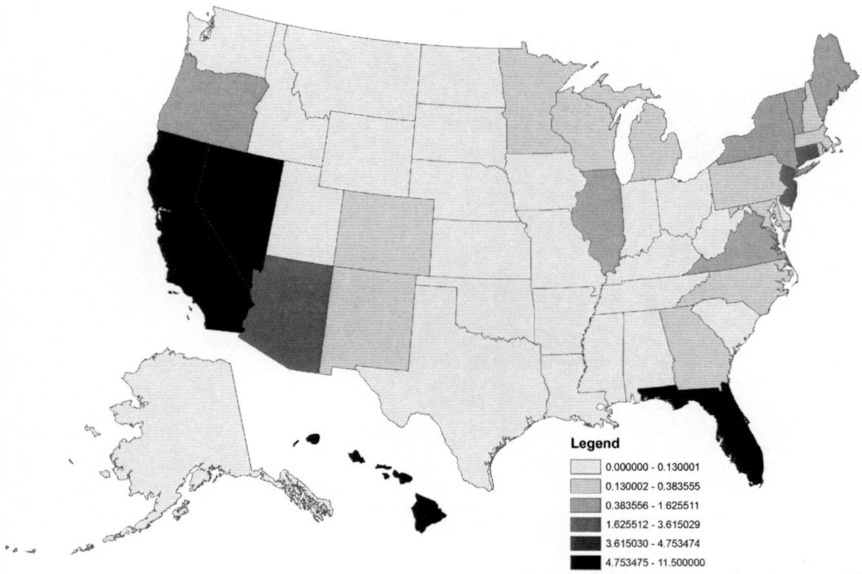

Source: Energy Information Agency (EIA)

Fig. 11.14 Solar location quotient, 2006

11.5.6 Wind Capacity

Wind capacity increased mainly in the Midwest just east of the Rocky Mountains, in the flat open plains stretching from Montana to Texas as illustrated in Figs. 11.11 and 11.12. The distribution of wind capacity in 1999 was confined to only a couple of states out of much of the west that had adequate wind resources. By 2006, the capacity appears to have increased in states with the greatest wind potential and more specifically in states where the average wind class is high enough to be considered for commercial wind energy production. In addition to the good wind resource availability there is also a greater likelihood that land is available for the construction of large scale wind farms without running into not-in-my-back-yard (NIMBY) problems. Total wind capacity is also low in Nevada and Arizona where overall wind class is suitable for wind but solar technologies may be more favored and therefore growth in wind capacity is brushed aside.

11.5.7 Solar Capacity

Figures 11.13 and 11.14 show solar capacity, measured in shipped square feet of solar panels, which saw the most growth in the southern, sunny states of California, Nevada, Arizona, Florida, and Texas. The growth appears to have continued to increase in those areas where the concentration of solar panels already existed. However, there was also an increased trend along the upper Midwest and New England states. It is unusual that states that have traditionally had little sunshine experience equal growth as those with higher resource potential. Therefore, the reduced solar insolation received in these states points to the technology being adequate regardless of the location, with the possible exception of Alaska. This map indicates that there is enough solar insolation to satisfy the solar needs of these states.

11.5.8 Chi-Square

The results of the chi-square analysis show a statistically significant association between: 1) the location quotient of wind and renewable portfolio standards $x^2(n = 50) = 11.377$, $p < .05$ and 2) the location quotient of solar and renewable portfolio standards $x^2(n = 50) = 11.925$, $p < .05$. Based on these results, the relationship between production and capacity for both solar and wind are statistically related to the presence of renewable portfolio standards.

Unfortunately, no statistical relationship was observed between the simple presence of generic "incentives". Consequently, even though consumer and producer incentives and average resource potential will have some effect, it is ultimately not the most important, nor the strongest relationship. Therefore, states that have enacted renewable portfolio standards (RPS) will have varying levels of growth but are more likely to have a larger overall capacity. In addition, each state with renewable portfolio standards had them in effect for an average of 5 years. Therefore, states with renewable portfolio standards have been requiring renewable energy production for many consecutive years and this may be the underlying reason for the existing relationship.

11.6 Limitations

There are some limitations to the study that must be addressed. First, no incentive data were collected between years for each state. After a review of incentives, at least one incentive was in effect for each year comprised in this study. However, there could be a time period within the year for which no incentive was available. For example, each incentive typically has a legislative end but are often revised or extended. Such revisions and extensions typically coincide with the end of the earlier versions but gaps could occur where an incentive is available, expires, then is reenacted later in the year. Therefore, this study does not account for any mid-year gaps in incentive availability.

This chapter also only measured the average resource potential for each state and did not account for any other factors that would keep wind turbines or solar panels from being setup. No assessment was made on available land, terrain, Not-In-My-Back-Yard (NIMBY) issues, climate, or weather patterns. The resource potential for wind also did not include areas off the coast. In addition, solar resource average was determined as if the collector was in a fixed position to the south, tilted at the same degree as their corresponding latitude which would ensure direct collection of solar radiation. Statistics for tracking plate collectors were not used since the average solar consumer would most likely have the panels fixed to their house roofs and the cost of a tracking system might also be cost prohibitive.

11.7 Conclusion

The growth and spread of wind and solar technologies are tied to a couple of key factors. First, the energy source must be plentiful enough to be captured and utilized. Second, the renewable energy technology must be developed enough so it is both stable and reliable. Last but not least, it is apparent that if an emerging technology is to gain early entrance into the market prior to it being economically competitive and superior to current technologies, then regulations must be instituted to force the early adoption of that technology. This study shows that governmental regulation is currently a key element in spurring the use of these more expensive and less diffused technologies. It shows that incentives and/or natural resource availability are not the main driving factors in wind and solar energy growth but are still important considerations. In addition, without a considerable increase in oil and fuel prices and environmental regulations renewable electricity generation will probably develop slowly in the United states, despite any increase in incentives at the state, local, or federal level or a normal increase in technology (Lin 2007; Menz 2005).

Therefore, to speed up the adoption of wind and solar energy production this study suggests that states must institute policies and regulations similar to renewable portfolio standards that require a percentage of electricity production from utilities to come from these technologies. If a concerted change towards the development of renewable technologies and energy policy is not promoted, most forecasts indicate that production from renewable technologies will remain a small percentage of total production and will have little short-term or long-term investment value (Wiser et al. 2004).

The findings of this chapter coincide with other studies that have evaluated RPS policies and have found similar results. Wiser (2005) noted that one study estimated that 47% of new wind capacity between 2001 and 2004 was the direct result of renewable portfolio standards. Menz (2005) noted that RPS used in combination with financial incentives seem to have been the most effective means of developing renewable energy technology and production. However, the adoption of RPS policies does not ensure an expected level of growth. Another study conducted by Wiser et al. (2004) found that even though a number of states experienced positive results from RPS policies, such as Texas, other states that had poorly designed theirs with limited applicability, inadequate economic planning or enforcement that ultimately translated into little to no effect on promoting renewable energy generation (Wiser et al. 2004).

Therefore, consistent demand for "green" electricity will be the only effective means to help develop renewable technologies such as wind and solar. As development increases, economies of scale will take hold making renewable technology cheaper or at least competitive with current technologies. However, the diffusion of renewable portfolio standards (RPS) or similar measures will need to be widespread in order for these technologies to have substantial growth. If the United States remains balkanized in its adoption of energy policy and regulations such as RPS, then growth will be modest until change is economically or politically necessary and not just a matter of public or political opinion.

References

Agnolucci, P. 2006. Use of economic instruments in the German renewable electricity policy. Energy Policy 34(18): 3538–3548.

American Wind Energy Associaton (AWEA). 2008. http://www.awea.org/ (last accessed August 13, 2008).

Armstrong, H. and J. Taylor. 1978. Regional Economic Policy and its Analysis. Oxford. Philip Allan.

Bezdek, R. and R. Wendling. 2007. A half century of US federal government energy incentives: value, distribution, and policy implications. International Journal of Global Energy Issues 27(1): 42–60.

Bird, L., M. Bolinger, T. Gagliano, R. Wiser, M. Brown and B. Parsons. 2005. Policies and market factor driving wind power development in the United States. Energy Policy 33(11): 1397–1407.

Calzonetti, F. 1981. Finding a Place for Energy: Siting Coal Conversion Facilities. Association of American Geographers. Washington, DC.

Chapman, K. 1974. Corporate systems in the United Kingdom petrochemical industry. Annals of the Association of American Geographers 64(1): 126–137.

Christopherson, Robert W. 2006. Elemental Geosystems. Upper Saddle River, NJ: Prentice Hall.

Clements, Donald W. 1977. Recent trends in the geography of coal. Annals of the Association of American Geographers 67(1), 109–125.

Darmstadter, J. 2000. The role of renewable resources in U.S. electricity generation – Experience and prospects. Resources for the Future: Washington, DC.

Deyette, J., S. Clemmer, and D. Donovan. 2003. Plugging in Renewable Energy: Grading the States. Cambridge, MA: Union of Concerned Scientists.

Dogniaux, R. 1994. Prediction of Solar Radiation in Areas with a Specific Microclimate. Commission of the European Communities. Kluwer Academic Publishers.

Database of State Incentives for Renewable & Efficiency (DSIRE). 2007. http://www.dsireusa.org (last accessed December 6, 2007).

Elmes, G. 1985. Modeling spatial interaction of utility coal in Pennsylvania. Annals of the Association of American Geographers 75(2): 212–226.

Elmes, G. 1996. Industrial restructuring and the United States coal-energy system, 1972–1990: Regulatory change, technological fixes, and corporate control. Annals of the Association of American Geographers 86(3), 507–529.

Energy Information Administration (EIA). 2005. Renewable Energy Annual, 2005. http://www.eia.doe.gov/cneaf/solar.renewables/page/rea_data/rea.pdf. (last 14 accessed June 2007).

Energy Information Administration (EIA). 2006b. Annual Energy Review 2006. http://www.eia.doe.gov/emeu/aer/contents.html. (last accessed 29 June 2007).

Energy Information Administration (EIA). 2004. Monthly Energy Review, April. http://tonto.eia.doe.gov/ FTPROOT/multifuel/mer/00350404.pdf (last accessed 6 December 2007).

Energy Information Administration (EIA). 2003. Annual Energy Outlook 2003 With Projections to 2025. http://tonto.eia.doe.gov/FTPROOT/multifuel/mer/00350404.pdf (last accessed 6 December 2007).

Gan, L., G. Eskeland and H. Kolshus. 2007. Green electricity market development: Lessons from Europe and the US. Energy Policy 35: 144–135.

Goldberg, M. 2000. Federal Energy Subsidies: Not all Technologies Are Created Equal. Renewable Energy Policy Project. No. 11.

Gouchoe, S., V. Everette and R. Haynes. 2002. Case Studies on the Effectiveness of State Financial Incentives for Renewable Energy. NREL/SR-620-32819, National Renewable Energy laboratory, Raleigh, NC.

Heiman, M. and B. Solomon. 2004. Power to the people: Electric utility restructuring and the commitment to renewable energy. Annals of the Association of American Geographers 94(1): 94–116.

Hudson, Ray and David Sadler. 1990. State policies and the changing geography of the coal industry in the United Kingdom in the 1980s and 1990s. Transactions of the Institute of British Geographers, New Series 15: 435–454.

Lancaster, R. and M. Berndt. 1984. Alternative energy development in the USA: The effectiveness of state government incentives. Energy Policy 12(2), 170–179.

Langniss, Ole and R. Wiser. 2003. The renewables portfolio standard in Texas: an early assessment. Energy Policy 31: 527–535.

Lins Jr., H. 1979. Energy development at Kenai, Alaska. Annals of the Association of American Geographers 69(2): 289–303.

Lewis, J. and R. Wiser. 2007. Fostering a renewable energy technology industry: An international comparison of wind industry policy support mechanisms. Energy Policy 35: 1844–1857.

Lydolph, Paul E. and Theodore Shabad. 1960. The oil and gas industries in the U.S.S.R. Annals of the Association of American Geographers 50(4): 461–486.

McVeigh, J., D. Burtraw, J. Darmstadter and K. Palmer. 2000. Winner, loser, or innocent victim? Has renewable energy performed as expected? Solar Energy 68(3): 237–255.
Menz, Fredric C. 2005. Green electricity policies in the United States: Case Study. Energy Policy 33: 2398–2410.
Menz, Fredric C. and S. Vachon. 2006. The effectiveness of different policy regimes for promoting wind power: Experiences from the states. Energy Policy 34: 1786–1796.
National Renewable Energy Laboratory (NREL). 2007. Solar Resources for the U.S. Department of Defense. http://www.nrel.gov/gis/femp.html (last accessed 29 June 2007).
National Renewable Energy Laboratory (NREL). 2007a. PV Solar Plate Radiation, Annual. (Flat Plate, Facing South, Latitude Tilt). http://www.nrel.gov/gis/images/us_pv_annual_may2004.jpg (last accessed 5 December 2007).
National Renewable Energy Laboratory (NREL) 2007b. Wind Power Today. http://www.nrel.gov/docs/fy06osti/39479.pdf. (last accessed 5 December,, 2007).
Nersesian, Roy L. 2006. Energy for the 21st Century: A Comprehensive Guide to Conventional and Alternative Sources. Armonk, NY: M.E. Sharpe, Inc.
Nogee, A, S. Clemmer, B. Paulos and B. Haddad. 1999. Powerful Solutions: 7 Ways to Switch America to Renewable Electricity. Union of Concerned Scientists.
Norberg-Bohm, V. 2000. Creating incentives for environmentally enhancing technological change: lessons from 30 years of U.S. energy technology policy. Technological Forecasting and Social Change 65(2): 125–148.
Ó hUallacháin, B. and N. Reid. 1991. The location and growth of business and professional services in American metropolitan areas, 1976–1986. Annals of the Association of American Geographers 81(2): 254–270.
Sawyer, S.W. 1983. State renewable energy tax incentives: monetary values, correlations, policy questions. Energy Policy 11(3): 272–277.
Sheskin, I. 1978. Alaskan Natural Gas: Which Route to Market? The Professional Geographer 30(2): 180–189.
Simon, Christopher A. 2007. Alternative Energy: Political, Economic, and Social Feasibility. Lanham, MD: Rowman & Littlefield Publishers, Inc.
Szyliowicz, Joseph S. and Bard E. O'Neill. 1975. Energy crisis and U.S. foreign policy. Praeger Publishers, New York.
Taylor, R. 2006. Solar Resource Assessment. National Renewable Energy Laboratory. http://www.eere.energy.gov/tribalenergy/solar.cfm (last accessed 29 June 2007).
Taylor, J. and P. VanDoren (2002). Evaluating the Case for Renewable Energy: Is Government Support Warranted? Cato Policy Analysis, No. 422.
U.S. Department of Energy (DOE). 2006. The History of Solar. http://www1.eere.energy.gov/solar/pdfs/solar_timeline.pdf (last accessed 15 June 2007).
U.S. Department of Energy (DOE).2005. History of Wind Energy. http://www1.eere.energy.gov/windandhydro/printable_versions/wind_history.html. (last accessed 15 June 2007).
Wiser, Ryan, Kevin Porter, and Robert Grace. 2004. Evaluating Experience with Renewables Portfolio Standards in the United States. Ernest Orlando Lawrence Berkeley National Laboratory. http://eetd.lbl.gov/ea/ems/reports/54439.pdf. (last accessed August 14, 2007).
Wiser, Ryan, Kevin Porter, and Robert Grace. 2005. An overview of policies driving wind power development in the West: State renewables portfolio standards and integrated resource plans. In Proceedings of National Wind Coordinating Committee Western Transmission Workshop III. Sacramento, California. February 1, 2005.

Chapter 12
Environmental and Social Influences on Historical County Creation in the United States

Karinne Rancie, Samuel M. Otterstrom, Jeffrey M. Sanders and Fredric J. Donaldson

Abstract This chapter uses a GIScience framework to explore the environmental and social factors that influenced the making of US counties from 1790 through 2000. Existing literature suggests that arbitrary political actions, carrying capacity, and time minimization explain the size and timing of county creation. These theories were examined using updated data and tested for their validity with a historical GIS and statistical analysis. The chapter found that political, sociological, and ecological factors need to be supplemented with geographical considerations, such as elevation, natural amenities, and proximity to like-sized counties to explain the evolution of county creation. This chapter is unique in that it represents a historical approach to GIScience and can be used to inform the geography of governance structures.

Keywords Historical GIS · County · Political boundaries · United States

12.1 Introduction

Counties are one of the most basic geographic units of government in the United States and have played a central role in the historical development of the nation's political landscape. From the colonial New England town in the northeast and Tidewater country of the south through to the westward expansion, counties have been the ubiquitous manifestation of federal and state governance throughout US history (Grant and Nixon 1968, p. 409). County governments, especially for the rural areas of the nation, are often the closest and most immediate public entity to which citizens have access. This political unit provides the opportunity for grassroots governance while also extending the control of state power (Grant and Nixon 1968, p. 410; Murphy 1970, p. 3).

S.M. Otterstrom (✉)
Department of Geography, Brigham Young University, Provo, UT, USA
e-mail: sam_otterstrom@byu.edu

J.D. Gatrell, R.R. Jensen (eds.), *Planning and Socioeconomic Applications*, Geotechnologies and the Environment 1, DOI 10.1007/978-1-4020-9642-6_12,

Furthermore, counties have important roles in both rural and urban areas. "Counties have had the image of being predominantly a rural instrument of government. This is because the county is the only form of local government in many rural areas" (Murphy 1970, p. 1). However, from the 1950s the United States has seen a significant movement toward urbanization, and the majority of counties are now comprised of both urban and rural areas with a great many consisting of an entirely urban population. This means that individual counties end up providing widely ranging services depending on the internal structures of their counties.

Although counties that have some governing authority are found across the US, with the possible exceptions of Connecticut and Rhode Island , the size and population characteristics of each county are far from standard. The average county size is approximately 961 sq. miles, with actual sizes ranging from the over 20,000 sq miles of San Bernardino County, CA to the 6 sq. miles of Fairfax City, VA. The demographics, or population density, of counties are just as varied ranging from the very dense New York County, NY with almost 70,000 people per sq. mile, to Loving County, Texas, which averages less than 1 person per sq. mile.

One quick glance at a map of the United States (Fig. 12.1) illustrates the variety in county size across the nation, as well as the noticeable increase in size as one travels west. However, the explanations for the differences in county size are not as

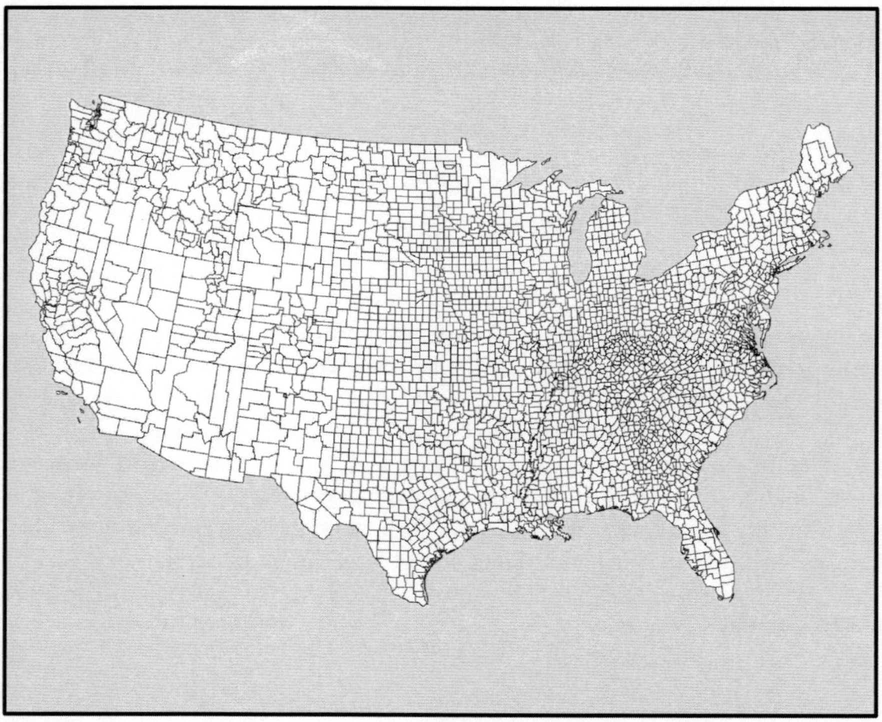

Fig. 12.1 Current county boundaries of the United States

obvious, which raises two basic questions. First, what historical and physical factors can account for such a large range of county sizes in the United States? Second, what are the best statistical methods to analyze the evolution of the nation from unsettled land to its current geographic pattern of county boundaries?

The purpose of this chapter is to systematically analyze the pattern of county subdivision and growth historically across the United States. With the benefit of updated county size and population density data, we use methods previously applied by other researchers to reevaluate and expand upon the theories that they have already proposed to explain US county creation. We also employ other spatial inquiry approaches to aid in our explanation. Through the process we will underscore the strengths and weaknesses of three main county creation theories to test their validity. Specifically, these three models relate to the concepts of carrying capacity, arbitrariness of political actions, and time-minimization of travel (Fox 1976; Schuessler 1974; Stephan 1971).

12.2 Geographies of US County Creation

In the last four decades there has been extensive work that examined the process of county creation within the United States. The studies have covered topics such as the proliferation of new counties, especially after the 1700s as the population grew and expanded west, and the decline in new county divisions in the wake of the spread of the automobile (Boulding 1968; Stinchcombe 1968; Murphy 1970; Grant and Nixon 1968). Additionally, numerous research articles have dealt with the relationship between county size and population density (people per square mile). The general, and perhaps relatively obvious, conclusion has been that counties with small areas usually have high densities, while those with large areas most often have lower relative population densities (Stephan 1971; 1972; Stephan and McMullin 1981; Fox 1976; Schuessler 1974).

However, when discussion comes to the topic of explaining why counties are the sizes they are, there has been no definitive solution or simplistic answer. For example, why do some counties vary radically in size even when they are in the same region and share similar physical characteristics? Why do some areas with similar population densities get divided into different sized units? In some cases it appears that the counties within certain regions have just random areas. Are there some geohistorical factors that can help account for these widely ranging governmental units? Because the different arguments used by previous authors to explain county sizes across the US landscape are not definitive and can contradict each other, this reevaluation is timely and necessary. To provide the necessary background we will review the work of Stephan, Fox, and Schuessler who propounded three main theories regarding the relationship between county size and population from different perspectives (Stephan 1971, 1972, 1976, 1979; Vining et al. 1979; Stephan and McMullin 1981; Fox 1976; Schuessler 1974).

Fox (1976) maintained that ecological factors in a region play a significant role in the size of counties. He argued that there is an inverse relationship between carrying capacity (number of individuals a region can support based on available resources) and area with lower population densities resulting in the large area/low carrying capacity counties. Additionally, he noted that natural features, such as islands, mountain ranges, or water sources, have had a considerable effect on county size by providing logical boundary lines. At the outset, Fox's theory seems to be helpful in explaining the general size differences between eastern and western counties: the arid western region had less of an ability to carry high population densities resulting in larger county areas, while the small counties in the East can be explained by more plentiful precipitation that could support a larger agricultural and/or urban populace.

Fox's argument is compelling in theory, yet in principle it does not appear to be applicable across the country. It is not just the carrying capacity of the county land itself, but whether the land could support more people because of outside inputs (such as in the case of emerging cities). For example, the high population density of the Los Angeles system of cities (or other urban areas in the west such as Phoenix) gives one the impression that the area has a high carrying capacity, and thus should have smaller country sizes. In reality, these arid western areas compensated for lack of water, which would normally determine carrying capacity, by tapping resources from other often distant areas so that an artificial carrying capacity level could be maintained (Cosby, 2004, From mission to megacity: the changing concentration of the Los Angeles City-System, unpublished MS thesis, Brigham Young University).

Schuessler (1974) argued that political boundaries were not drawn by sociological processes, but by "arbitrary political action." An inverse correlation exists between county area and population-density because it appears obvious that county areas will be large when there is a low population density, and there will be smaller areas where the population density is higher. This inverse relationship can be, and is, created by political reasoning rather than being underpinned on social necessity. However, historical evidence points to the fact that many counties were at some point divided after their initial creation for various reasons, not always related to population growth. Schuessler's theory of arbitrariness, then, does not seem to provide an adequate reason for these subsequent divisions and their re-formations into new government units.

The explanations given by both Fox and Schuessler have been offered largely in response to the body of work published by G. Edward Stephan. Stephan believed that the inverse relationship between county area and population density was primarily influenced by sociological processes. In the first of Stephan's publications related to county creation, he proposed the "time-minimization theory" of population segmentation (Stephan 1971). According to this line of reasoning, counties have been structured, restructured, and segmented to reduce the time spent in their maintenance and to meet the needs of the local society. Based on prior work of Durkheim (1933), Boulding (1968), and Stinchcombe (1968), Stephan provided the explanation that new county growth resulted from the significant movement of people from the heavily populated east coast to the western frontier. According to the theory, as the population spread westward, political centers (county seats) were located where people could participate in the sociological functions of the county seat without an unreasonable travel burden. This was true during the time period when local transportation methods were stable (horse and wagon) (see Murphy 1970, p. 2). Furthermore, the line of reasoning continues that the segmental growth of new counties ceased around 1930 because the rise of the automobile and highways effectively made travel constraints less relevant.

Stephan's subsequent publications focused on developing and justifying the time-minimization theory by empirical application to both United States and international situations. Stephan (1972) applied the theory to 98 countries, and the political units within each country, and found that in the vast majority of the cases, 94 out of 98, there was a negative relationship between political unit area and population density. Stephan formulated, and termed the "world regression line" relating area to population density as a line with a slope of $-2/3$ (1972, p. 367). In a related study in Great Britain, a subsequent overhaul of the county system by the UK government brought the county area/population-density relationship to within the expected vicinity of the world-regression line (Massey and Stephan 1977), which further confirmed to the authors that the regression line was a reasonable way to estimate county sizes via population density.

To further test the general applicability of the time-minimization theory, Stephan and Tedrow (1977) applied it to the relationship between urban area and population – an important effort as most of the previous U.S. analysis had been performed using rural counties – and found that an urban area's size generally varied by population density as expected. Mycielski and Trzeciakowski (1963) earlier work also supported Stephan's time-minimization theory as they showed that the location of service stations was more related to convenience and accessibility than distance. Additionally, Palmer (1973) illustrated mathematically that service locations were located to minimize travel while Stephan and McMullin (1981) sought to reinforce time-minimization theory by applying it to the historical realities of county creation in the United States, but they did not use a complete set of historical county boundary maps.

Although much has been published on the regression line between area and population density, there has been very little spatial analysis using actual historical data when testing the correlation between the incidences of area and population density. The oversight is surprising, especially as Stephan and McMullin (1981) asserted that spatial distribution relative to political centers, not size, makes subdivisions necessary. The regression analysis implied that the variables of area and population density were independent. However, we believe that the assumption of independence is not met with the county area and population data because of spatial autocorrelation (Anselin 1990). In this case, spatial autocorrelation of population density occurs because settlement intensity across the country is not randomly located, but tends to be clustered in certain parts of the continent (see also Griffith et al. 2003).

Stephan and McMullin (1981) focused on rural counties rather than on all types of areas because the time-minimization theory does not lend itself well to urban areas. They argued that limiting their analysis to rural areas was workable because the population in the US was predominantly rural during much of the county creation period. We now recognize the need to consider how urbanization has also impacted the development of counties. Although the majority of the US population lived in rural areas in 1790, the bulk is presently found in urban counties.

The theoretical works of Fox, Schuessler, and Stephan have introduced three related, yet competing, explanations of the current spatial configuration of counties across the United States. What this current body of scholarship lacks, however, is a statistical analysis of a comprehensive historical database that includes county boundaries, population, and other data that would lend credence to or discredit these contrasting theories. According to Stephan (1971), the ideal study of the size-density relationship of counties should consider all county boundary data and all population data for the entire history of the United States from its formation to the present (which he did not attempt). Our study will therefore work to close the gap between theory and historical reality through the use of historical digital county boundary maps coupled with population figures and other variables that we analyze using a variety of statistical techniques.

12.3 Analytical Approach

This section underscores the data gathering and analysis process that we followed in examining the theories of Fox (1976), Schuessler (1974), and Stephan (1971) to better understand the forces that shaped the sizes and distribution of counties in the United States. Although we spend time examining each of the three, we focus more attention on Stephan's theory due to the more extensive nature of his work.

Beginning in 1790 and continuing with subsequent decades, the data from the US Census allow a relatively accurate accounting of county populations, so we set the chronological framework for the study as 1790 through 2000. County area estimates were not included in the censuses until 1900. However, the Historical United States County Boundary Files [HUSCO 1790–1999] (Earle et al. 1999) has area estimates for all US counties for each decade from 1790 to 1890 along with the census area figures from 1900 onward. We therefore used these two data sets, the census populations and historical county area estimates, to analyze the county size-density relationship.

The analysis of Fox's argument, that the size of counties is determined by carrying capacity, is performed by using area and elevation to estimate a county's ability to provide for viable economic activities for its inhabitant. We consider elevation to be functioning as a "natural boundary" that constrains population growth at higher altitudes. We should expect to see a negative relationship between a county's area and its carrying capacity, and thus the higher a county's elevation the larger the size should be. A result that does not fit this paradigm would require further explanation. The relationship will be analyzed at the national and regional (divisional) levels (using the regions designated by the US Census Bureau (Fig. 12.2)). Additionally, a spatial autocorrelation exploration of county area (global and local Moran's I) and a cluster analysis of county elevation and area are performed to examine the

Fig. 12.2 Regions of the United States (US census divisions)

geographic effects of spatial proximity. If Fox's natural boundary theory is valid, then the spatial autocorrelation and cluster statistics should show a nonrandom pattern.

Schuessler claimed that counties were created on the arbitrary whim of politicians, not sociological processes. A test of correlation performed at the county level between size and population density of counties, if found to be significant, will lead to clarification of the area to population relationship including whether sociological variables validate his hypothesis (urban/rural majority, urban populations, and industry dependence as shown through thematic county maps). If Schuessler's theory is correct, then we would expect to see only a small level of, or no correlation between county size and population density.

We assumed that county data from the US Census Bureau have normal distributions, as they are generally full population censuses rather than random samples. With that characteristic in mind, the Pearson's r was the primary statistic used to determine the level of correlation between population density and county size, rather than the non-parametric Spearman's r. However, a comparison of results between the two tests will allow a greater range of inferences to be made from the results.

Adding to the correlation tests that we performed to test Schuessler's theory, we followed Stephan's linear regression techniques to compute the "world regression line" comparing county size and population density. Our advantage is that we used updated and complete data of all US rural counties' populations and densities for all decades from 1790 to 1990. The linear regression computed to analyze the relationship between population density and county area follows the form: $\log A = K + b(\log D)$ where A is rural county area, K is the intercept, D is rural population density, and b is the slope. The slope should be negative, according to Stephan's theory, meaning that as the log of population density decreases, the area of the county increases. Stephan and McMullin (1981), in his test of the time-minimization theory argued that the tremendous variability in the county area and population density warranted using the logarithmic transformation. These regression results are reported and compared with Stephan's, and with 1990 population density Global Moran's I spatial autocorrelation statistics.

County size and population density correlation is used because Stephan identified density as a determinant of county creation and size. In the case of low correlation coefficients, the data are displayed on a scatter plot to illustrate any linear association between the variables. Cases with population densities over 200 people per square mile were excluded in the analysis to minimize urban outliers, enabling a clearer view of the relationship between area of counties and population density.

In summary, the intent of the chapter is to provide analysis of the three main theories behind county creation using the statistics along with a series of thematic maps to demonstrate the geographical relationships between county size and a number of other variables. These thematic maps include: natural amenities, industry dependence, relief, private/Federal land ownership, elevation zones, land use, and urban/rural dichotomies.

12.4 Fox's Theory of Carrying Capacity

Correlation analysis shows that there is a significant relationship between the size of a county and its average elevation (Table 12.1). At the national, aggregate level, the correlation coefficient is 0.501 with a significance of 0.01. This means that higher elevations generally equate with larger county sizes, and by extension higher elevations mean lower carrying capacities per acre. Correlation coefficients are often deemed to be significant, even when they are relatively low, due to the large number of cases involved. However, it is noteworthy that the size and elevation correlation showed the highest coefficient of any pair of variables that we compared at the national level.

Table 12.1 2000 Carrying capacity correlation – Pearsons' r

	Area/Elev (county)	Area/water coverage (state)	(Ave water area – sq miles) (state)	(Ave area – sq miles) (county)	(Ave elevation – ft) (county)
Nation	**0.501**	0.202	–	–	–
New England	**0.288**	*0.879*	1531	941	633
Middle Atlantic	**0.526**	0.636	3295	665	783
E. N. Central	**0.369**	**0.942**	11570	558	820
W. N. Central	**0.373**	0.558	1778	822	1686
S. Atlantic	**−0.122**	*0.657*	3358	476	576
E. S. Central	**−0.387**	**0.983**	1201	491	738
W. S. Central	**0.426**	0.396	4351	919	937
Mountain	*−0.1*	−0.221	936	3054	5413
Pacific	**0.388**	0.734	4959	2399	2633

Note: Bold = significant at 0.01, Italics = significant at 0.05

We also computed the same correlation coefficients at the regional level. Positive correlation was again the result in the majority of cases. However, three regions had significant negative correlation: the Mountain, East South Central, and the South Atlantic regions. The Mountain West has generally high elevation and the largest counties, while quite small counties and low elevations are found in the East South Central and South Atlantic regions. These negative results weaken Fox's argument. If the carrying capacity/natural boundaries relationship to county size was valid, one would expect a significant and positive correlation between elevation and county size in the Mountain region. However, the results show that this region's correlation had the lowest level of significance along with a small negative correlation coefficient although it has the highest average elevations and the largest average size of counties of all the regions.

The Moran's I statistic gives a different perspective on the relationship between place and county size. It showed significant spatial autocorrelation of similarly sized counties with the Global Moran's I of area for 1990 county boundaries being 0.694 with a 0.001 p value, which is a fairly high positive spatial autocorrelation level. Additionally, Local Moran's I statistics indicated significant spatial clusters of large counties adjacent to other large in the West and small counties in the Southeast (Fig. 12.3).

This indicates the importance of local geography and regional historical conditions in these areas that influenced the creation of many large or small counties within close proximity. What the Local Moran's map does not emphasize is the

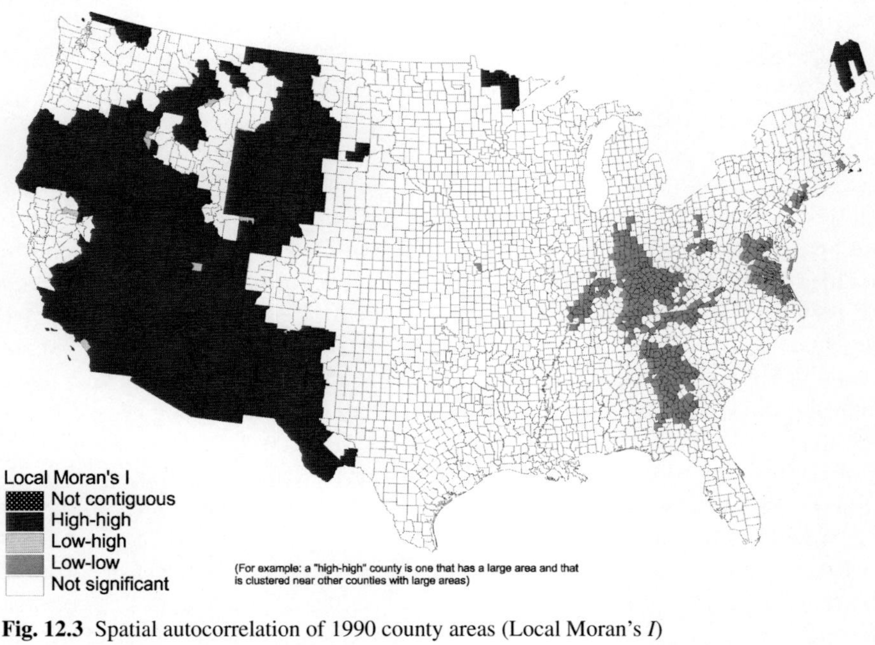

Fig. 12.3 Spatial autocorrelation of 1990 county areas (Local Moran's *I*)

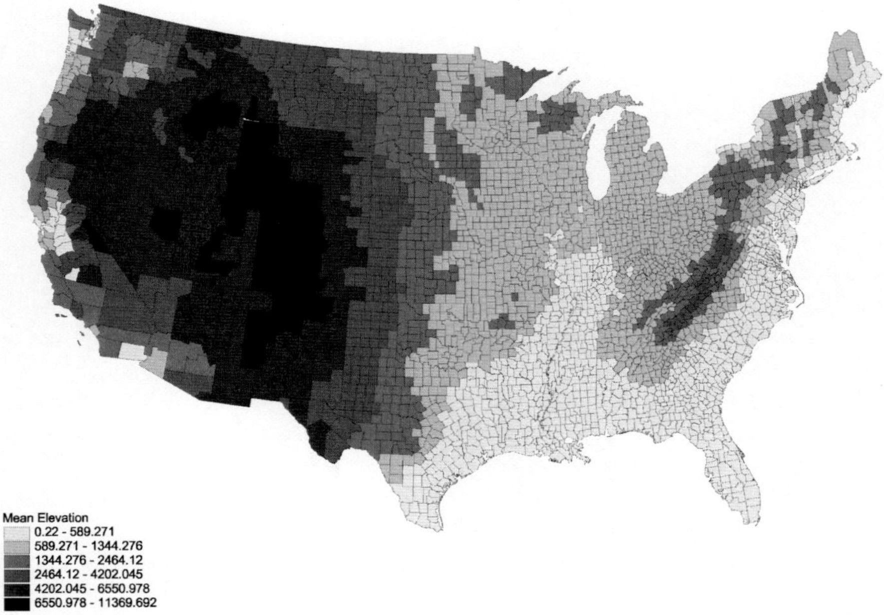

Fig. 12.4 Mean elevation of US counties

preponderance of middle sized counties from the Great Plains to the East that also cluster together.

Elevation and county area are quite related. First, a quick visual survey of the map of mean elevations of counties shows some correlations in county areas (Fig. 12.4).

More specifically, the map illustrates the k-means cluster analysis of a combination of county area and mean county elevation shows very strong spatial clustering of counties with similar sizes and elevations (Fig. 12.5 and Table 12.2).

It shows areas delineated as follows: low areas and elevations covering most of the eastern half of the country as well as part of the coastal counties of the West (cluster 5); moderate elevations and areas specifically showing the Appalachians and bordering counties of the Rockies (cluster 4); high elevation and smaller size counties of the Rocky Mountains (cluster 2), as well as a handful of counties in the mountains of California and Nevada; medium sized counties, mostly in the Mountain region (cluster 7), surrounding in good part mountainous cluster 2 but not having as high of elevations; semi-arid counties in the Mountain Region with fairly high elevations and large sizes (cluster 1); four very large, mostly high desert counties (cluster 6); miscellaneous Western counties (many are coastal or on national borders) having medium to lower mean elevations but fairly large sizes (cluster 3); and very densely urbanized counties that area mostly small in size (omitted from the cluster analysis).

These results show that counties with similar size and population density do cluster around specific elevation zones, meaning that that there is a relationship between elevation and county size. Therefore, Fox's theory regarding the effect of natural

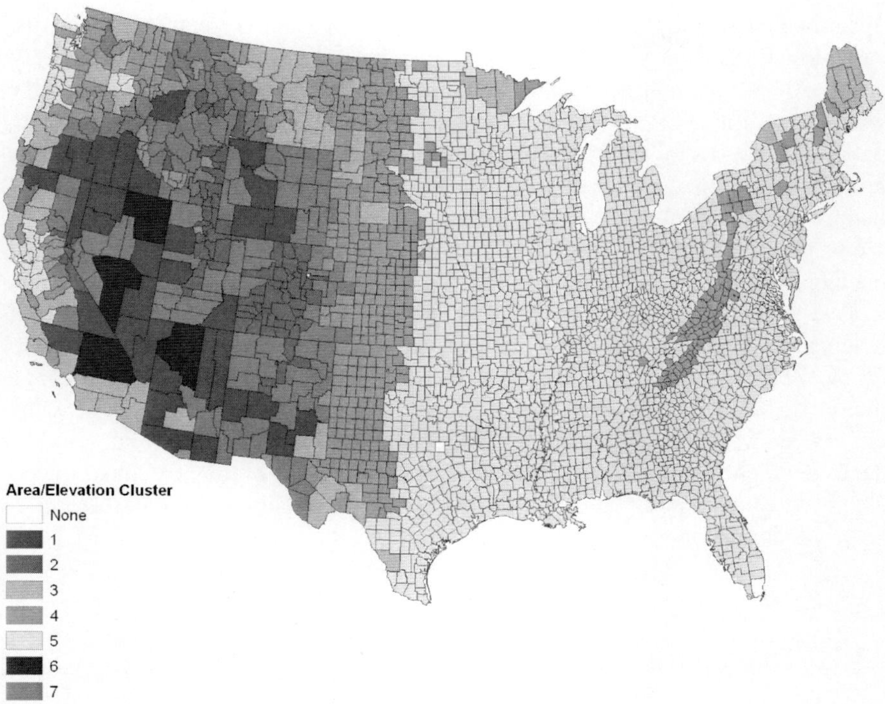

Fig. 12.5 County area/elevation clusters (from k-means cluster analysis)

Table 12.2 County area/elevation cluster centers

Cluster	1	2	3	4	5	6	7
Area square miles	8300.6	1324.8	4079.0	1090.0	605.4	18481.7	3293.2
Mean elevation feet	**5086.3**	**7595.8**	**2300.2**	**2660.4**	**728.8**	**5130.1**	**5436.6**
Number of counties	33	82	55	435	2315	4	121

Note: Bold = significant to at-least .05

boundaries on county size does seem to have some validity when differing elevations are translated into varying levels of potential carrying capacity. However, other factors besides elevation that could relate to carrying capacity must be considered in order to better validate Fox's premise.

12.5 Schuessler's Theory of Political Arbitrariness

12.5.1 Correlation

The initial correlation coefficients between county size and population density using Pearson's r were quite weak (> −0.2 in all decades), although the results did show the negative correlation that was expected (using all counties). Interestingly, the

Spearman's r statistics showed stronger negative levels of correlation between the variables. In the first year, 1790, Spearman's r resulted in a very negative coefficient (−0.742), with the remaining years showed results between −0.46 and −0.66, which were still quite strong for the large number of counties we analyzed. While we are most interested in the Pearson's r statistic, since the variables are at the interval level, the Spearman's statistic provides a helpful relative comparison that shows that there is some inverse relationship between density and county area, and it also seems to account for extremes in county area or population density in its stronger readings.

As the analysis at the aggregate level did not produce higher correlation coefficients, we computed the same statistic using regional level clusters of counties (Table 12.3). We found that although the relationship was generally significant at the 0.01 level, the coefficients were again low. The coefficient was highest within the first few years of settlement for each region, and was then found to substantially decrease in correlation within 50 to 60 years – providing an obvious link between county size and population density at initial county creation but not necessarily as local settlement intensity changed over time.

12.5.2 Qualitative Analysis

The aim of the qualitative analysis is to determine what makes the regions of the Middle Atlantic, East North Central, South Atlantic, and East South Central have higher correlation coefficients of county size and population density than the other regions, and why the Pacific region did not show significant correlation levels. This will be accomplished through a qualitative analysis of two sets of mapped variables. These characteristics are urban/rural county designations (urban population/population density), and industry dependence by county.

The map of urban/rural county designations (Fig. 12.6) shows counties that are defined as Metropolitan, Nonmetropolitan – "micro" (near metropolitan center), and Nonmetropolitan – non-core (rural, not near metropolitan center). It can easily be seen that in those regions where correlation is mostly significant, counties are predominantly metropolitan – a significant proportion of the US population resides within these high population density regions. Additionally, although the majority of the counties are metropolitan, there are also significant numbers of non-core counties and micro counties adjacent to metropolitan areas. All three county types appear to be interspersed through the regions in small clusters.

In the Pacific region, where correlation is not significant, there are large areas of metropolitan counties with smaller pockets of non-core and micro counties. In the central portion of the country where there was also less significant correlation, the pattern is one of mostly non-core rural counties with small pockets of metropolitan and micropolitan counties. We deduce that there could have been a greater reliance on using local population density levels to determine county boundaries in the East where there was higher settlement intensity across the landscape because of plentiful

Table 12.3 County correlations of area and population density by region (Pearson's r) – 1790 to 2000

Year	New Eng.	Mid-Atlantic	E. N. Central	W. N. Central	S. Atlantic	E.S. Central	W.S. Central	Mtn	Pacific
1790	**−0.532**	−0.193	–	–	**−0.463**	−0.424	–	–	–
1800	*−0.272*	*−0.19*	−0.478	–	**−0.376**	**−0.551**	–	–	–
1810	*−0.249*	**−0.246**	*−0.279*	*−0.864*	**−0.343**	**−0.443**	−0.358	–	–
1820	*−0.234*	**−0.263**	**−0.253**	*−0.474*	**−0.28**	**−0.423**	*−0.366*	–	–
1830	−0.191	**−0.241**	**−0.196**	*−0.326*	**−0.169**	**−0.533**	*−0.267*	–	–
1840	−0.201	**−0.217**	**−0.143**	**−0.34**	**−0.155**	**−0.525**	*−0.193*	–	–
1850	−0.177	**−0.21**	**−0.185**	−0.121	**−0.188**	**−0.466**	−0.033	−0.389	−0.249
1860	−0.144	**−0.211**	**−0.248**	**−0.136**	**−0.112**	**−0.367**	−0.073	**−0.426**	−0.042
1870	−0.134	**−0.225**	**−0.187**	−0.043	*−0.106*	**−0.307**	−0.058	**−0.29**	−0.079
1880	−0.128	**−0.229**	**−0.181**	*−0.08*	*−0.104*	**−0.282**	*−0.089*	**−0.335**	−0.089
1890	−0.126	**−0.233**	**−0.124**	*−0.093*	**−0.106**	**−0.215**	−0.06	**−0.288**	−0.091
1900	−0.136	**−0.245**	*−0.089*	−0.065	**−0.111**	**−0.187**	*−0.097*	**−0.299**	−0.093
1910	−0.136	**−0.252**	−0.071	*−0.076*	*−0.088*	**−0.13**	**−0.123**	−0.093	−0.093
1920	−0.138	**−0.23**	−0.056	*−0.08*	**−0.11**	*−0.103*	**−0.132**	−0.078	−0.093
1930	−0.139	**−0.273**	−0.043	*−0.082*	**−0.123**	−0.067	**−0.124**	−0.077	−0.092
1940	−0.138	**−0.274**	−0.04	*−0.083*	**−0.125**	−0.06	**−0.119**	−0.077	−0.094
1950	−0.142	**−0.277**	−0.042	*−0.084*	**−0.119**	−0.032	*−0.107*	−0.078	−0.097
1960	−0.157	**−0.303**	−0.047	*−0.093*	**−0.198**	−0.019	*−0.102*	−0.081	−0.103
1970	−0.171	**−0.318**	−0.05	**−0.103**	**−0.196**	−0.02	*−0.098*	−0.091	−0.112
1980	−0.184	**−0.316**	−0.055	**−0.118**	**−0.193**	−0.023	*−0.098*	*−0.106*	−0.12
1990	−0.019	**−0.313**	−0.058	**−0.125**	**−0.188**	−0.02	*−0.09*	*−0.113*	−0.122
2000	−0.19	**−0.317**	−0.059	**−0.134**	**−0.183**	−0.026	*−0.082*	*−0.115*	−0.126

Note: Bold = significant at 0.01, italics = significant at 0.05.

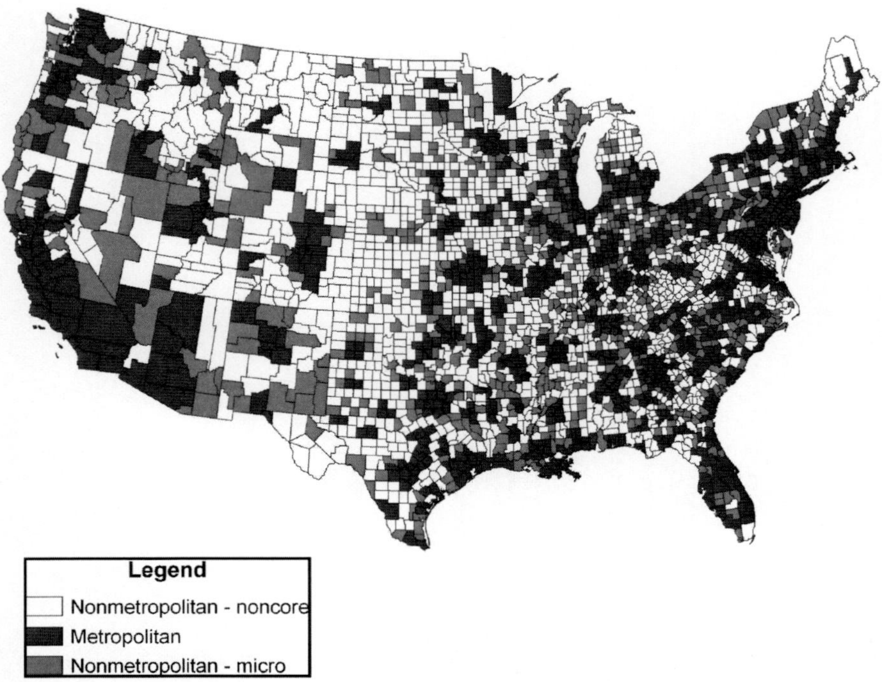

Fig. 12.6 Metropolitan, micropolitan, and rural county (US census designations)

rainfall, good access, and arable land. Those factors translated into the spatial variety of county types in the eastern regions today. In contrast, the larger one-type clusters in the Pacific (mostly metropolitan and high population density) or the rural predominance in the Midwest are probably partly a product of local economic and urban conditions that have diminished the underlying relationship of area to population density.

Typology codes, which show the most important industry in each county of the country, can also be analyzed for relational patterns to county size. In the eastern regions where the correlation between county size and population density is significant, there is a great reliance on manufacturing, together with service and non-specialized (finance, etc.) industries. In the Pacific region, non-specialized economies are dominant, with a smattering of other types of counties (especially services, government, and manufacturing). In the remaining regions, economic dependence is generally on farming and Federal/State Government, along with non-specialized and mining industries.

County typologies hint at a relationship between a county's industry and county size and population density. In areas where there is a high percentage of non-specialized industry, especially in the Pacific region, population density and county size do not match up very well. On the other hand, eastern manufacturing counties in the East could help indicate where there is a higher correlation between population density and county size. Perhaps, initial pods of manufacturing establishments and employees living in close proximity gave rise to counties that had fairly well-matched county size/density arrangement.

The correlation and qualitative analyses support the proposition that some local characteristics and processes helped determine county boundaries, and that new counties did not just come about by arbitrary political action. Therefore, the incidence of significant correlations and their relationship to settlement category and economic typology emphasizes that these current variables relate to historical conditions that influenced the delineation of county boundaries in the country.

12.6 Stephan's Theory of Time-Minimization Regression

The relationship between the variables of rural county area and population density were next analyzed using linear regression. The resultant averaged regression equation was Log A = K − 0.387(log D). The −0.387 coefficient substantially differed from the negative two-thirds (−0.66) relationship that was posited by Stephan (see Fig. 12.7 and Table 12.4). However, for the most part our results visibly mirror Stephan's time-minimization test results, albeit at a lower level. This means that the population density and size variables of counties were not as strong in their inverse relationship as believed by Stephan (see also Table 12.3). Stephan noted that size-

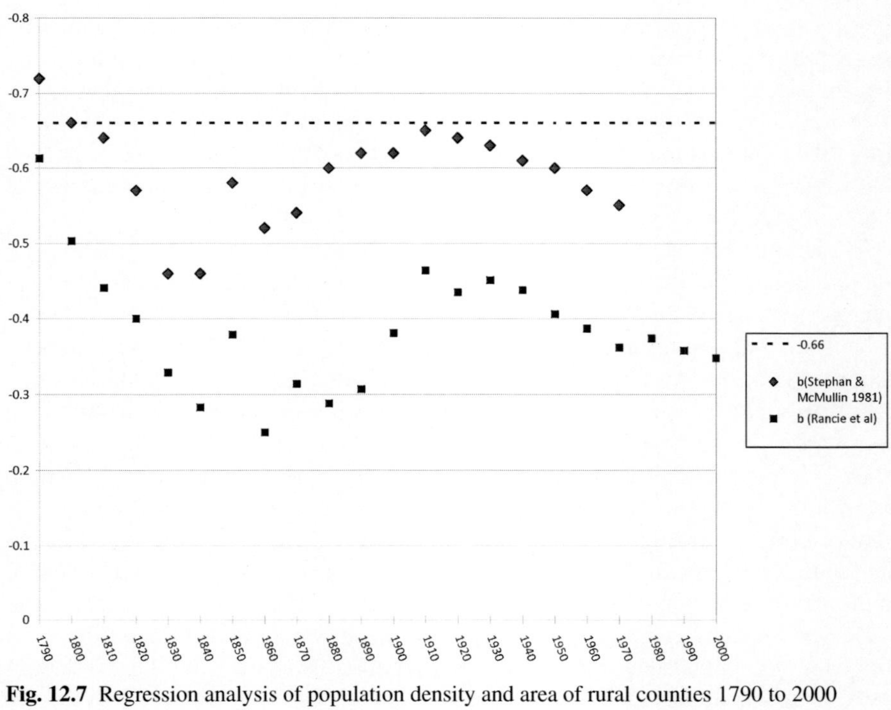

Fig. 12.7 Regression analysis of population density and area of rural counties 1790 to 2000

Table 12.4 Regression results of population density and area of rural counties 1790–2000

Census year	No. of units (rural counties)	R Square	b	B (Stephan & McMullin 1981)	Positive
1790	271	0.542	−0.613	−0.72	0.72
1800	408	0.462	−0.503	−0.66	0.66
1810	562	0.482	−0.441	−0.64	0.64
1820	744	0.459	−0.4	−0.57	0.57
1830	962	0.393	−0.329	−0.46	0.46
1840	1244	0.289	−0.283	−0.46	0.46
1850	1569	0.484	−0.379	−0.58	0.58
1860	2020	0.333	−0.25	−0.52	0.52
1870	2208	0.448	−0.314	−0.54	0.54
1880	2455	0.402	−0.288	−0.6	0.6
1890	2623	0.406	−0.307	−0.62	0.62
1900	2623	0.462	−0.381	−0.62	0.62
1910	2722	0.338	−0.464	−0.65	0.65
1920	2751	0.464	−0.435	−0.64	0.64
1930	2725	0.446	−0.451	−0.63	0.63
1940	2723	0.288	−0.438	−0.61	0.61
1950	2650	0.247	−0.406	−0.6	0.6
1960	2569	0.23	−0.387	−0.57	0.57
1970	2488	0.212	−0.362	−0.55	0.55
1980	2336	0.318	−0.374		0
1990	2295	0.305	−0.358		0
2000	2217	0.296	−0.348		0

density hypothesis would hold only as long as transportation remained constant, suggesting that the relationship between U.S. county size and population density began to weaken in the 1920s with the proliferation of automobiles (Stephan 1971). Our results show consistently lowed R square values beginning in 1940, which might be an indication of how changing population densities in counties that have had mostly fixed boundaries after 1910 have diminished the current applicability of Stephan's theory.

Stephan and McMullin (1981) discussed the slope trend in historical context, but the differences between their slope and our study's results require additional explanation. The different units of analysis employed may explain some of this variance: counties in our study, states and number of county seats in Stephan's. Additionally, since we considered all rural counties compared with Stephan and McMullin's less inclusive study, it is not unreasonable that our statistics would be different. Although our regression coefficients are closer to 0 than Stephan and McMullin's our pattern shows an important relationship connected with the growth of the nation. The drop from 1790 through to the 1860s and 1870s followed by a rise to 1910 reflects the rapid frontier expansion of the country westward with new counties being platted and settled very quickly (with lower densities at first that translated into more neutral regression coefficients). This was followed by the end of frontier settlement around 1910 when county creation had greatly slowed and new settlement began to focus more on urban areas then rural areas. This supports Otterstrom and Earle's (2002) 1910 final date of the closure of the frontier.

12.7 County Creation Timing

Stephan (1971) states that segmental growth, the addition of counties in a state, ended around 1930, ostensibly caused by the growing use of the automobile and mass transportation. However, out of the nine regions (divisions) delineated by the U.S. Census Bureau (Fig. 12.2 and Table 12.5), only the Mountain and South Atlantic had significant (>2%) segmental growth rates until 1930, the other seven regions having already virtually ceased adding new counties: the Pacific and West South Central by 1920, the West North Central by 1910, and the remainder before 1900. Furthermore, the New England and Middle Atlantic regions had ceased adding a significant number of counties many years earlier by 1860 (Table 12.3). Thus the cessation in new county creation appears to have had more to do with new settlement progression and population densities than with twentieth-century transportation innovations.

12.8 Qualitative Analysis Using Thematic Maps

We use this qualitative analysis to consider a variety of geographical factors that could help explain county size and population density that have not garnered much attention from previous theorists. The list of possible geographical considerations is extensive (elevation zones, water consumption, water area, land cover, etc.). However, most of the variables can be combined or separated to form some logical categories: level of natural amenities, land ownership, and land use. These human land relationships are also related to levels of precipitation, elevation, and access to sunlight. As in the analysis on Schuessler's theory, we use the correlation between county size and population density for the comparative basis of the analysis. Specifically, we examine why there are higher rates of correlation in the Middle Atlantic, East North Central, South Atlantic, and East South Central regions between county size and population density than the remaining regions, and why the Pacific region shows very weak correlation in these variables.

We first explored the United States Department of Agriculture natural amenities scale (that is based on a number of climatic and other environmental factors). Areas that have a combination of greater relative amounts of warm winters, winter sun, temperate summers, low summer humidity, topographic variation, and water areas have the highest amenity levels. Figure 12.8 shows that in the regions of significant correlations of county areas and population density there is a lesser degree of natural amenities than in those regions of less correlative significance. In areas with fewer scenic attractions or with a less desirable climatic environment, population growth and higher densities themselves have been seen as an acceptable use of land and would therefore encourage smaller county sizes.

Of course, there are areas in which there is high population density, large county size and high amenities, as well as areas of low amenities and low correlation significance. Therefore there is certainly not a clear delineable relationship among

Table 12.5 Growth in the number of counties by region 1790–1990 (total number and percentage change from previous decade)

Year	New England		Middle Atlantic		East North Central		West North Central		South Atlantic		East South Central		West South Central		Mountain		Pacific	
1790	41		49				172		16									
1800	47	14.6	76	55.1	9			224	30.2	62	287.5							
1810	51	8.5	98	28.9	46	411.1	6		248	10.7	103	66.1	21					
1820	55	7.8	114	16.3	120	160.9	15	150	263	6	159	54.5	32	52.4				
1830	58	5.4	121	6.1	201	67.5	32	113.3	313	19	207	30.2	53	65.6				
1840	62	6.9	130	7.4	306	52.2	80	150	354	13.1	267	29	77	45.3				
1850	64	3.2	142	9.2	351	14.7	158	97.5	392	10.7	290	8.6	176	128.6	14		34	
1860	**67**	**4.7**	**146**	**2.8**	401	14.2	349	120.9	459	17.1	305	5.2	254	44.3	31	121.4	82	141.2
1870	67	0	147	0.7	410	2.2	415	18.9	471	2.6	330	8.2	271	6.7	97	212.9	94	14.6
1880	67	0	148	0.7	421	2.7	533	28.4	484	2.8	**351**	**6.4**	357	31.7	119	22.7	100	6.4
1890	67	0	148	0	**432**	**2.6**	613	15	495	2.3	356	1.4	386	8.1	164	37.8	118	18
1900	67	0	149	0.7	435	0.7	588	−4.1	552	11.5	356	0	408	5.7	190	15.8	126	6.8
1910	67	0	149	0	436	0.2	**610**	**3.7**	517	−6.3	361	1.4	456	11.8	207	8.9	130	3.2
1920	67	0	150	0.7	436	0	619	1.5	538	4.1	364	0.8	**469**	**2.8**	269	29.9	**133**	**2.3**
1930	67	0	150	0	436	0	621	0.3	**557**	**3.5**	364	0	470	0.2	**280**	**4.1**	133	0
1940	67	0	150	0	436	0	620	−0.2	555	−0.4	364	0	470	0	280	0	133	0
1950	67	0	150	0	436	0	619	−0.2	555	0	364	0	470	0	281	0.4	133	0
1960	67	0	150	0	436	0.2	619	0	562	1.3	364	0	470	0	279	−0.7	133	0
1970	67	0	150	0	437	0	619	0	561	−0.2	364	0	470	0	279	0	133	0
1980	67	0	150	0	437	0	618	−0.2	561	0	364	0	470	0	279	0	133	0
1990	67	0	150	0	437	0	618	0	561	0	364	0	470	0	281	0.7	133	0
2000	67	0	150	0	437	0	618	0	561	0	364	0	470	0	280	−0.4	133	0

Note: Bold = significant to at-least .05

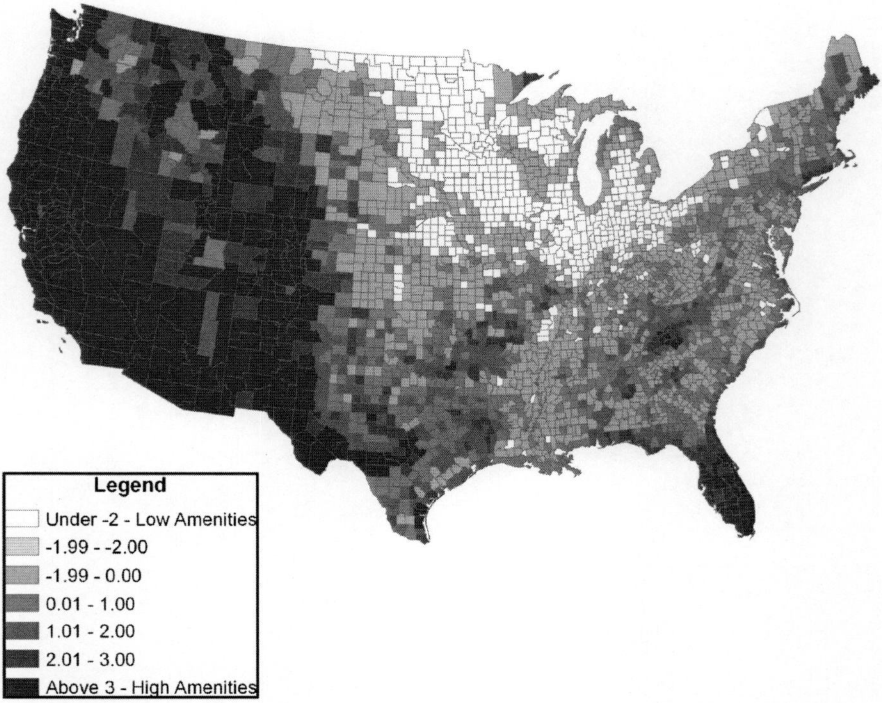

Fig. 12.8 Natural amenities levels of United States counties (Source: Economic research service – USDA)

county size, population density and amenities. However, the amenities map does illustrate a strong visual connection between county size and amenities, with many large western counties showing high amenities scores and much of the small county areas of the East (and especially the upper Midwest) having low amenities. The line that divides counties of higher amenity levels from those of lower amenities begins in western Texas and goes north through New Mexico, Colorado, Wyoming and central Montana. This boundary is also an appropriate divider between large county regions and small county regions.

Considering land use, there is also a definite boundary between 'rainy' areas of the east and mountain (arid) areas of the West, and the type of land uses that exist there. Although the relationship is complex, counties that have more rain and arable land often have smaller areas and higher population densities. Those counties with mountainous topography, less rain, and/or dominated by grazing, show less of a relationship between area and population density, and often have larger areas and low population densities. Therefore, it appears from this short analysis that levels of precipitation and topographical factors that influence land uses are factors that have been instrumental in affecting county areas.

There is also a similar relationship between type of land ownership and level of natural amenities (Fig. 12.9).

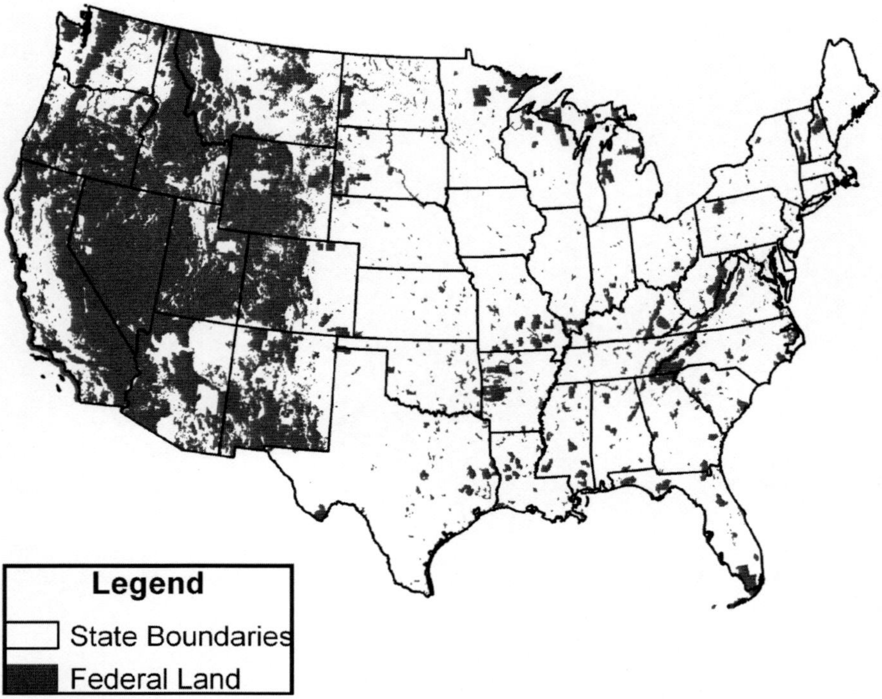

Fig. 12.9 Federal lands of the United States (Source: Nationalatlas.gov (2005))

Federally owned lands can be found mainly in the mountainous regions of the West along with smaller pockets including Appalachia, while privately owned lands dominate the East where there is a significant correlation between county size and population density. Much federal land is now considered high amenity, but was originally thought of having less productive potential. This federal ownership is correlated with large county sizes and lower densities (partly because there is less land open for settlement) in many areas such as Nevada, Arizona, and Utah. On the other hand, private lands in the East are usually of greater population density and smaller because they have been settled more fully.

From this qualitative analysis we have seen that those counties in which natural amenities abound, and those that are under greater control or protection from the federal government, are more likely to have less population density and larger county sizes. In lands where there is a lower level of natural amenities, private ownership abounds and with more land open to development, higher population densities and smaller county sizes have resulted.

12.9 Conclusion

This study has shown that the county creation theories of Fox and Stephan have some validity and do contribute to understanding the relationship between county size and population density to varying degrees. However there was less evidence of validity in regards to Schuessler's theory. Even though significant correlations between population density and area exist at times in certain areas, we have shown that sociological and physical variables factor into the process of county creation more than political arbitrariness alone. Fox's related theory of carrying capacity did reveal the importance of the ability of a county's area to support its population initially, with the strongest point being that topographical characteristics, especially elevation, play a role in determining county boundaries and area.

Stephan noted that the size-density hypothesis would hold only as long as transportation remained constant, meaning that it would have weakened in the 1920s with the national spread of the automobile. The results of the regression analysis confirmed a softening in the relationship beginning around 1930, but also that the relationship was actually weaker than Stephan posited. We also must note that the high levels of autocorrelation impacted the validity of the regression analysis (in 1990 the Global Moran's I statistic of spatial autocorrelation for population density was a highly significant 0.413 (p value of 0.001)). Additionally, the correlation analysis illustrated a stronger relationship among county size, original settlement date and population density than the impact of the advent of widespread automobile use.

Our statistical analysis has shown that county creation and subdivision is a process that, in addition to political, sociological and ecological factors, has a geographical component not properly measured singly by ordinary regression analysis. Any study that attempts to provide an explanation for county growth must take into account the spatial environment in which the county division process occurs. Doing so would yield more valuable insights into the spatial dimensions and growth of not only counties, but also other political and administrative units. Further study could focus on integrating statistical analysis with the geographical variables used in the qualitative analysis to make a more unified theoretical explanation of the county creation process.

Testing the three county creation theories and adding insights provided by qualitatively analyzing other factors, has offered unique glimpses of the varied historical and ecological aspects that have impacted county creation. The distinctive qualities of this data have required careful analysis and proper methods. Historical GIS has proved indispensable in this regard as previous studies made assumptions and proffered conclusions without considering all historical county boundaries in time or space. Thus, we have been able to show that a number of geographical relationships such as elevation, population density, and proximity to like-sized counties have helped broaden our understanding of how the United States local political geography evolved into its peculiar present configuration.

Acknowledgments We gratefully acknowledge the late Carville Earle for suggesting the potential for revisiting the subject of county creation in the United States. Additionally, we appreciate the work of Joel Warren, Laurie Weisler, Natalie Lester, and Erica Trone for their mapping and editing assistance. We also acknowledge the financial support of various entities at Brigham Young University.

References

Anselin, L (1990) Spatial dependence and spatial structural instability in applied regression analysis. J Reg Sci 30(2):185–207
Boulding K (1968) The city as an element in the international system. Daedalus 97:1111–1123
Durkheim, E (1933) Emile Durkheim on the division of labor in society. Macmillan, New York
Earle C, Otterstrom SM, Heppen J (1999) The historical United States county boundary files (HUSCO). Geoscience Publications, Baton Rouge, Louisiana
Fox JG (1976) Tests of the size-density hypothesis: a caution. Am Sociol Review 41:567–569
Grant DR, Nixon HC (1968) State and local government in America. Allyn and Bacon, Inc., Boston
Griffith D, Wong D, and Whitfield T (2003) Exploring relationships between the global and regional measures of spatial autocorrelation. J Reg Sci 43(4):683–710
Massey D, Stephan GE (1977) The size-density hypothesis in Great Britain: analysis of a deviant case. Demograph 14(3):351–361
Murphy, TP (1970) Metropolitics and the urban county. Washington National Press, Inc., Washington DC
Mycielski J, Trzeciakowski W (1963) Optimization of the size and location of service stations. J Reg Sci 5(1):59–68.
Nationalatlas.gov (2005) http://nationalatlas.gov/ – 1 October 2008
Otterstrom SM, Earle C (2002) The settlement of the United States from 1790–1990: divergent rates of growth and the end of the frontier. J Interdiscip Hist. 33(1):59–85.
Palmer DS (1973) The placing of service points to minimize travel. Oper Res Q 24(1):121–123
Schuessler K (1974) Analysis of ratio variables: opportunities and pitfalls. Am J Sociol 80(2): 379–396
Stephan GE (1971) Variation in county size: A theory of segmental growth. Am Sociol Review 36:451–461
Stephan GE (1972) International tests of the size-density hypothesis. Am Sociol Review 37: 365–368
Stephan GE (1976) Reply to Fox. Am Sociol Review 41:569–570
Stephan GE (1979) Political subdivisions and populations density. Sci 205:219–220
Stephan GE, Tedrow LM (1977) A theory of time-minimization: the relationship between urban area and population. Pac Sociol Review 20(1):105–112
Stephan GE, McMullin DR (1981) The historical distribution of county seats in the United States: a review, critique, and test of time-minimization theory. Am Sociol Review 46(6):907–917
Stinchcombe AL (1968) Constructing social theories. Harcourt, Brace & World, New York
US Census Bureau www.census.gov – 1 Oct 2008
Vining DR, Yang CH, Yeh ST (1979) Political subdivisions and populations density. Sci 205:219

Chapter 13
Local Government Use of GIS in Comprehensive Planning

Mark W. Patterson and Nancy Hoalst-Pullen

Abstract The use of GIS by local governments for planning applications is increasingly becoming commonplace. In some cases, GIS departments have evolved into what Roger Tomlinson calls an enterprise-wide system, in which agencies easily share data among themselves. This chapter explores recent trends in the use of GIS for planning. We examine organizational structure, infrastructure, data sharing and data analyses. Specifically, we look at the use of GIS for compiling a Comprehensive Plan in Cobb County, Georgia and highlight models used by the County for green space preservation and industrial assessment. We conclude that the county's decentralized organizational GIS structure facilitates data collection, maintenance and updating, and the enterprise-wide system infrastructure allows for easy data sharing among agencies and the public. These characteristics were vital in helping to assemble the county's latest comprehensive plan.

Keywords Comprehensive planning · Future land use mapping (FLUM) · Green space · Georgia (US)

13.1 Introduction

Geographic information systems (GIS) are becoming increasingly more commonplace in local governments. Intrinsically, GIS is a tool that helps local governments understand what is in their jurisdiction, how it is functioning, and what the future will look like when geographic data and models are examined within the context of public opinion and government needs. This is particularly true with regard to community planning, and in the creation and implementation of a comprehensive plan.

A comprehensive plan (also known as a master plan or general plan) is central to the decisions of land use and planning for local government, as it is the one of the few planning documents (and for some places in the United States, the only planning document) that collates informationfrom a range of agencies and programs to

N. Hoalst-Pullen (✉)
Department of Geography and Anthropology, Kennesaw State University, Kennesaw, GA, USA
e-mail: npullen@Kennesaw.edu

J.D. Gatrell, R.R. Jensen (eds.), *Planning and Socioeconomic Applications*,
Geotechnologies and the Environment 1, DOI 10.1007/978-1-4020-9642-6_13,

account for the current and future activities and needs of the private and public lands within the jurisdiction of the local government (Kelly and Becker 2000). Essentially, a comprehensive plan is a framework on to which decisions regarding land use and zoning, environmental quality, open and green space, transportation and public safety, utilities and infrastructure, housing and public facilities, historic preservation, and social planning occur (Kelly and Becker 2000; Nedović-Budić 2000). This list, however, is not exhaustive. All these decisions, however, directly relate back to the future of the community, its citizens, and its stakeholders.

According to Kelly and Becker (2000), the key to the success of any comprehensive plan is threefold: (1) the plan must be geographically comprehensive; (2) at a minimum, the plan must be comprehensive in all local issues concerning the future of the land within the jurisdiction of the local government; and (3) the plan must be long-range to project land use and other aforementioned variables several years (or decades) into the future. We argue that there is a fourth aspect of importance: the necessity for local governments to find ways of disseminating this information to the public (citizens and stakeholders, as well as within the government itself) before, during and after the creation of the plan.

The need for this dissemination of data, both among government agencies and with the public at large, arguably has made the use and application of geographical information systems in local government ubiquitous (Ceccato and Snickars 2000; O'Looney 2000). In many cases, GIS plays a central role in the data gathering and analyses of information used in creating comprehensive plans. In addition, GIS helps to implement, monitor, and update spatial data (Waite 2005) relating to the comprehensive plan and its related policies. Furthermore, GIS and geospatial tools give citizens and stakeholders access to accurate and up-to-date information, especially with the recent trend of providing such data in an available and/or interactive format via the Internet (Bresnahan 2005).

13.2 Trends in GIS for Local Government

As noted by Chan and Williamson (1999) and others, spatial data infrastructures (SDIs) using GIS technologies are being developed at the local political-administrative levels for the support of social, economic, and environmental decision-making. An SDI is a network of servers that provides access to spatial data for a large number of users for a wide range of applications, including comprehensive planning. In a survey done nearly a decade ago, Warnecke et al. (1998) found that over 70% of large local governments, defined as counties with populations exceeding 50,000 and cities exceeding 25,000, use GIS technologies. Bresnahan (2005) indicates that this growth has increased, with less populated counties and smaller cities now having similar infrastructure and GIS capabilities. Sarkar (2003) reports that a US Department of Interior 2004 survey found that of the 1,156 local governments that responded, 97% with populations larger that 100,000, 88% with-populations between 50,000 and 100,000 and 56% of smaller jurisdictions with less than 50,000 use GIS applications and mapping technologies. In addition, counties use GIS at a slightly higher rate than cities (72 to 64%).

How GIS is integrated into local governments varies, and differs in terms of its original initiation and implementation phases (Budić and Godschalk 1994; Rogers 1983; Zaltman et al. 1973). For example, current infrastructure, management, resources, and uses all affect the level of GIS integration within a government (Waite 2005). The integration of GIS technology in local government is complex, dynamic, and is situation and/or resource-based. For example, Budić and Godschalk (1994) note from their study of four North Carolina government agencies that the adoption and success of GIS was based on how GIS was incorporated (i.e. mandated by the organization versus adapted from individuals or subunits, selective versus holistic, formal versus informal, etc.), by the utilization of GIS in performing tasks, and by the management of the GIS technologies, specifically infrastructure, resources and personnel training.

Similarly, discusses three environments in the development and use of GIS within a local government: first, the desktop environment with stand alone software and individual user; second, the development (developer) environment with the importance being new applications and software development; and third, the server environment with web services to support an enterprise system. With the need for data accessibility of greater importance, the migration toward the server environment (e.g. interactive web based mapping) is occurring within many local governments.

O'Looney (2000) notes the perception that GIS is a necessary and important decision-making tool for local governments is growing. However, a potential issue with the organizational structure of GIS technologies is the propensity for redundancies in data collection and storage, particularly when GIS databases are created with little coordination among various levels of government and agencies (Nedović-Budić et al. 2004; Nedović-Budić and Pinto 1999a,b). With the growing need for data collection and data management, a result has been the creation of decentralized GIS departments within the government structure, with multi-participatory or shared GIS emphasis. In this situation, each department requiring GIS has its own GIS personnel, who are responsible for compiling, maintaining and updating pertinent data. These departments receive assistance from a GIS department in terms of data standardization, storage, and infrastructure (Waite 2005). While such an organizational structure may be more costly compared to a centralized system in which a GIS department oversees *all* spatial data needs, the benefits of more up-to-date and non-redundant geospatial data may outweigh the additional resource burden of trained personnel in each department (Nedović-Budić and Pinto 1999a). The decentralized GIS department model is fundamental to maintaining systems and data resources between departments (as our case study shows) and provides different levels of GIS data and applications that can be accessed and upgraded as related to various needs of local government department and agencies.

Regardless of whether local governments adopt a centralized or decentralized organizational structure, a trend over time in the US in terms of data sharing has been from department specific systems to multi-departmental agreements, to multi agency, cross-departmental and statewide sharing agreements (Croswell 1994, in Nedović-Budić and Pinto 1999a). More recently, the role of the internet has further increased the accessibility and availability of data sharing to involve participants outside the organizations or agencies.

From an earlier study by Budić (1994), GIS-related users reported improvements in communication of information, data availability, accessibility, and data accuracy for local governments. Other factors influencing planning decisions included political support, staffing, experienced staff, and the number and type of GIS applications. Of these noted points, the use of GIS via experienced staff and the number and type of GIS applications used is critical. GIS engagement is not uniform among all users or departments at the local government level, and thus, the use and importance of advanced GIS applications in urban planning may not be occurring. A trend noted in literature (Petit and Pullar 1999) is the ongoing need for more advanced spatial analysis. While the adoption of GIS for parcel mapping and asset management has been well documented (Dueker and DeLacy 1990), spatial analysis and modeling is of greater importance, especially when considering planning is multifunctional and necessary to help determine the transformations and interplay between human and environmental conditions and needs. The use of GIS extensions (e.g. Liu and Zhu 2004), spatial or geostatistical analysis (e.g. Brody et al. 2006), risk assessment analysis (e.g. Flax et al. 2002) as well as web-based geospatial applications (e.g. Drummond and French 2008), among other applications, can greatly improve the goals established by the local government through its comprehensive plan.

The literature documents various uses of GIS for comprehensive planning, though the majority of these are academic in nature and not specifically results provided by the local government. Two examples of the use of GIS in comprehensive planning are the preservation of biodiversity and green space, and the development of industrial growth, including "smart growth" i.e. limiting urban sprawl and promoting urban redevelopment as well as changes in residential patterns (e.g. exurban sprawl). Considering the urban landscape is often a choice between preserving biodiversity and green space and allowing for industrial and economic growth, GIS plays an important role in providing geographic knowledge on the use and function of the land. Comprehensive plans using GIS can spatially identify, model, and project possible scenarios concerning the economic and environmental integrity of the landscape (Sandström et al. 2006; Brody 2003).

Overall, GIS is important to local governments in terms of urban planning via comprehensive planning as it can take seemingly unrelated information or data and synthesize them in a geographic manner by way of maps and map images. The use of GIS has been multifaceted, and includes aspects of data gathering, data analysis, policy making, implementation and monitoring. As these maps allow for constant upgrading due to their digital format, their role in comprehensive planning are dynamic and provide a means to relay geographic information to local government officials, policy makers, and citizens. The following section examines the role and use of GIS with regard to Cobb County, Georgia's 2030 comprehensive plan.

13.3 Cobb County, Georgia

Cobb County, Georgia, is located to the northwest of the city of Atlanta (Fig. 13.1). With an estimated 2005 population of 628,988 people, Cobb County is the 4th most populous county in Georgia, and with an area of over 340 square miles, is 81st in size (out of 159 counties). Interstate 75 divides the county into east Cobb and west Cobb. East Cobb is "built-out", while development of west Cobb

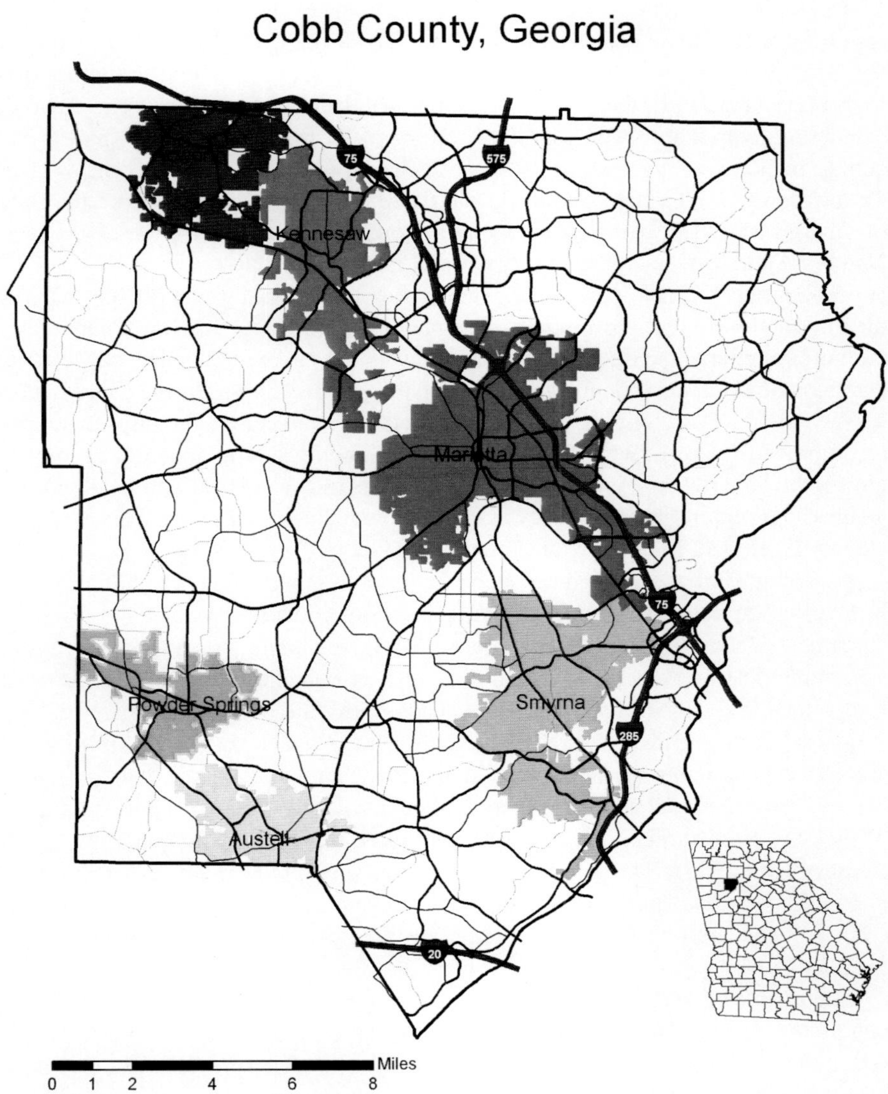

Fig. 13.1 Location of Cobb County, Georgia

is not far behind (Cobb Country Development Agency 2006, p. 34). Development in east Cobb is characterized as infill, with mixed use projects becoming more common. Most new developments in west Cobb are single family residential, with some abutting industrial zoned parcels. With the rapid expansion of residential developments, the county potentially faces a scarcity of both industrial land and green space. Both land uses are of importance to the county, as industrial zoned land is viewed as critical to maintain economic competitiveness, while green space is acknowledged as enhancing the overall quality of life for its citizens.

The rapid development seen in Cobb County makes the management of natural resources within the county complex. There are several public agencies at various levels of government charged with managing land and water resources in Cobb County. In addition to the six municipalities within the county (Acworth, Austell, Kennesaw, Marietta, Powder Springs, and Smyrna), there are tracts of land in the county owned and/or managed by other agencies, including county agencies (e.g. Cobb County Solid Waste Authority, and Cobb County Water Authority), the state of Georgia (e.g. Department of Transportation) and the federal government (e.g. National Park Service and the Army Corp of Engineers).

Cobb County is also part of the Atlanta Regional Commission (ARC), a coordinating body of local governments designed to oversee regional issues including, but not limited to: planning, transportation, water resources, and housing. The ARC encompasses 10 metro counties surrounding the city of Atlanta. As the coordinating agency, the ARC requires all local governments (municipal and county) to submit a comprehensive plan, which outlines the long range planning goals for each government. The ARC approves or requires revisions for each plan submitted to ensure a consistent regional based plan. The ARC approved comprehensive plan for Cobb County is entitled, Mapping Our Future: 2030 Comprehensive Plan.

As the title of the comprehensive plan implies, maps play a significant role in helping the county government and stakeholders make decisions and arrive at agreements. Thus, GIS has a significant role in the planning process, as maps serve as communication devices, as visual aids, and as a medium to illustrate potential future land use scenarios. The use of maps helps the county strive toward the vision set forth by the comprehensive plan. These goals are emphasized by the comprehensive plan's mission to envision Cobb County as a place that "promotes positive community characteristics and an area's sense-of-place" while simultaneously, providing "direction and a road map for the future of the community" (Cobb Country Community Development Agency 2006, p. 1). The comprehensive plan is compiled by the county's Community Development Agency, which is the lead county agency responsible for coordinating public meetings, as well as writing the plan.

The next section of the chapter examines how GIS was used to assemble the comprehensive plan, as well address some of the issues faced by Cobb County's GIS and planning departments. We begin with an overview of the comprehensive plan requirements as they pertain to GIS and mapping in general. Next, we examine GIS within the county, focusing on organizational structure and data collection issues. This is followed by providing examples of data analyses conducted by the GIS department. We conclude with a discussion on how GIS aided the planning process.

13.4 Comprehensive Plan

Under Georgia law, the Georgia Planning Act 634 requires each government to prepare and implement a comprehensive plan. The goal of the Planning Act is to create a framework that facilitates a coordinated, regional approach to planning, rather than having many disparate, and potentially conflicting comprehensive plans. Each plan, therefore, is subject to review by the respective Regional Development Center, which in Cobb County's case is the Atlanta Regional Commission. The Cobb County 2030 comprehensive plan, essentially an updated version of the 1995–2015 plan, accounts for the ever-changing social, economic, and environmental climate of the county. The plan states, "[I]n the year 2030, Cobb is recognized as Georgia's most complete community, a place that combines the best of urban, suburban, and rural life to appeal to a broad spectrum of people." (Community Development Agency 2006, p. 3). Thus, the overarching goal of the comprehensive plan is to provide guidelines to help direct the county to achieve a vision of what it wants to be in the future.

Considering the comprehensive plan document is essentially a community plan document, stakeholder and citizen participation is critical, particularly from the public. A series of public workshops were held as part of the plan's Public Participation Program (PPP). The goals of the PPP were to inform the public, receive input and allow the public to review and comment on proposed policies in the document. As with any planning document that includes public opinion,, the plan readily acknowledged that conflict and differences of opinion are inevitable. While nowhere was it explicitly stated that GIS be used, GIS did play a vital role in the PPP. Maps created by the county's GIS department provided the public with current information on a variety of issues such as land use, transportation, wetlands, population density, and cultural resources. Overall, the use of GIS in making maps helped to minimize any conflict based on geographical misinformation or lack of knowledge, and allowed for more effective communication among interested parties. For example, the use of GIS was important with regard to the green space acquisition portion of the comprehensive plan, which is discussed in detail below.

The Future Land Use Map (FLUM) (see Fig. 13.2) is the centerpiece of the comprehensive plan and is the official land use document of the comprehensive plan with respect to growth policy. The resulting future land use map is the culmination of the policies contained in the document making it the official "future development map", as requested by the Georgia Department of Community Affairs in their "Standards and Procedures for Local Comprehensive Planning..." (Community Development Agency 2006, p. 18).

The current FLUM is based on the maps and analyses discussed below, in addition to a host of other spatial and non-spatial data layers, such as transportation, administrative boundaries, water bodies, and tax records. The FLUM is a live map,

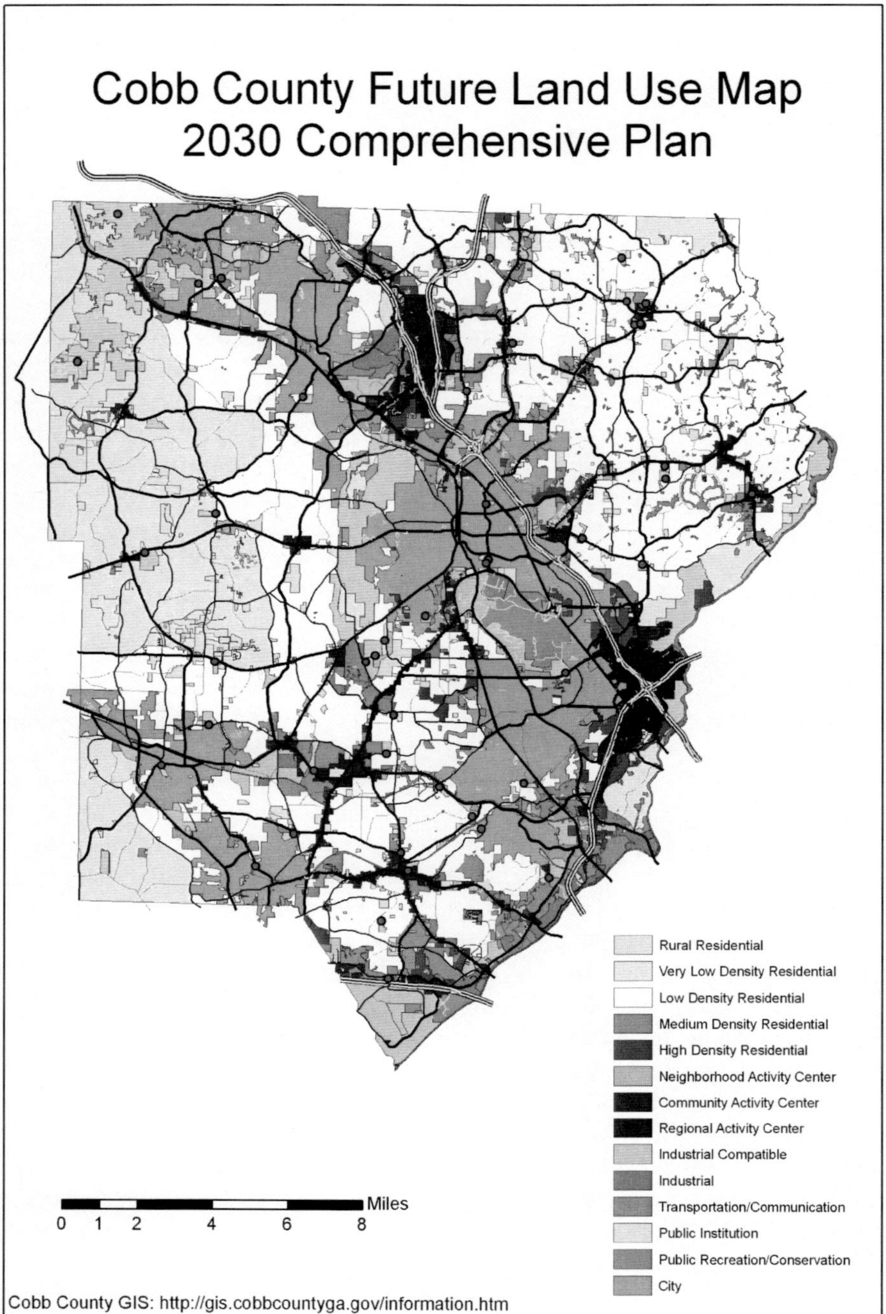

Fig. 13.2 Future land use map (FLUM)

with possibility of amendments and edits being made to reflect the changes in the preferences of the public and local elected officials.

13.5 Cobb County GIS

GIS use within the county can be characterized as one with a modified decentralized organizational structure. The GIS department oversees the daily and long term operations of GIS hardware and software, establishes data standards and is the primary warehouser of most spatial and attributes data. The department also coordinates data collection and acquisition for projects involving departments or other governments. One example is the annually acquired aerial imagery. While this is used for several purposes by several departments, it is the GIS dept that is charged with its acquisition and dissemination. In addition to aerial photography, there are a number of other departments who use GIS, including, transportation, planning, emergency services, and water. Each of these departments has its own GIS specialist, who is responsible for maintaining and updating the data (spatial and non-spatial) pertaining to his/her department. Departments can read other departments' data, but cannot edit or modify their data. While this modified decentralized organizational structure requires each department using GIS to have its own GIS personnel with expertise in the department's domain, it does offer the advantage of allowing departments to control their data and to update their data as needed. One of the positive outcomes of this arrangement is that data are relatively up-to-date in each department. The arrangement of this infrastructure is what Tomlinson (2003) calls an enterprise-wide system.

Like most public sector GIS agencies, Cobb County GIS has faced the issue of data sharing with the public. Currently, the county has a website (www.cobbgis.org) which allows users to create maps. Users have eight map layers from which to choose (e.g. roads, zoning, water, aerial photos) with each layer having several sublayers that users can toggle on and off. Standard tools such as zoom in/out, pan and identify are included. Users also have the option to save the map they create as a JPG or a PDF. In addition, users can create and save spatial bookmarks for subsequent sessions. And though the website is user friendly and provides an interactive tool to examine spatial data, the actual raw data are not downloadable.

Data from Cobb County available for download at no cost are limited, and include county boundary, city boundaries, transportation features, and the spatial data files needed to create the Future Land Use map. The county assumes users have access to GIS software to view the files. Imagery and spatial data available at cost include aerial orthophotos (1inch to 100 feet), elevation data (2 foot contours) and certain planimetric data.

13.6 Data Analysis

While GIS can produce maps to facilitate discussion on planning issues, the real power of GIS lies in its ability to perform spatial analysis. Cobb County GIS uses GIS to conduct data analysis ranging from simple buffers to slope analysis to economic analysis. This section explores some analyses have been conducted and incorporated into the comprehensive plan. The analyses range from basic spatial data analysis such as buffering to more advanced modeling analyses used for green space preservation, and for assessing industrial development and redevelopment within the county. Results from these analyses are used in constructing the Future Land Use Map of the comprehensive plan.

In terms of simple spatial mapping, Cobb County uses data from the US Fish and Wildlife Service to designate and protect wetlands in the county. The US Fish and Wildlife Service, one of the government agencies with jurisdiction in Cobb County, has created a map of wetlands in the county that Cobb County incorporates into their Future Land Use Map. Wetlands, as defined by the Clean Water Act (Section 404), are supposedly guaranteed protection from incompatible industrial and residential development. Likewise any undeveloped land intersecting a wetland is excluded from future potential residential development.

The County also uses GIS to implement the Metropolitan River Protection Act (MRPA), which is designed to protect rivers by limiting pollution from land disturbance activities. In response, a 2000 foot buffer (corridor) surrounds the Chattahoochee River in Cobb County. Properties which are currently available for development inside the buffer are identified and a flagged in the Future Land Use map as limited use only, with industrial and industrial compatible zoned uses highly discouraged. Applications for development inside the 2000 foot buffer are subject to a series of reviews by both the county and the ARC to ensure proposed activities are in line with the MRPA.

In addition to mapping wetlands and river corridors, GIS is used to compute the area of green space in the county, identify existing green space for protection and help identify areas in the county where more green space should be set aside. The comprehensive plan recognizes the importance and value of green space in the county. Cobb County is committed to protecting 20% of its total area as green space. This commitment is particularly difficult given the amount and rate of development in the county. To help uphold this commitment, the county adopted the Trust for Public Lands' Greenprint model, which examines several variables that together produce a map identifying high-priority areas where green space preservation or acquisition is needed.

Conservation priorities used in the Greenprint model include the following: (1) park gaps and needs; (2) undeveloped land; (3) existing support and infrastructure; (4) hydrology; (5) educational facilities; and (6) historic sites. Maps are created for each priority and then all are overlaid to determine areas with the highest needs for preservation.

The first variable, park gap and needs, is computed by analyzing demographic (census) data and performing a spatial query. Areas with a high percentage of the population under 18 years and with low household income levels (less than $25K) are weighted higher by the model than other areas in the county. Areas greater than 0.25 miles from an existing park are also weighted higher. The second variable, undeveloped lands are those parcels with at least 75% of vegetation in a natural state and containing one or no structures. With existing support and infrastructure, the idea is to ensure continuity of trails. This is computed by selecting parcels that are less than 300 feet from existing trails. The forth variable, hydrology, includes factors such as proximity to wetlands, floodplains, streams and water bodies, and highly permeable soil. Parcels located closer to these features are assigned a higher weighting. Parcels with public facilities such as schools, libraries and senior centers are given a higher weighting than those without. Finally, parcels with historic structures on the National Registry of Historic Places are assigned a higher weighting than those parcels without such structures.

The model allows for different weightings to be assign to each of the above conservation priorities when producing the final map. Figure 13.3 shows a green space preservation priority map with all six priorities assigned an equal weighting, with the darker gray areas having the highest priority for preservation of green space.

Environmental issues are not the only application of GIS for which the county uses GIS. With respect to economic and community development, the county strives to have an appropriate amount of industrial zoning. Cobb County GIS has developed an industrial assessment model which is used to identify potential locations where industry can be developed. This information in turn aids the county when in trying to encourage industrial firms to locate in the county.

The model examines the attribute and spatial data of the county's parcels, as well as spatial relationships of these parcels and other features. Attribute-based criteria used in the model include: zoning, assessed property value, market readiness, development potential and area specific incentives. The variable zoning is used to exclude from analysis all parcels not zoned Industrial or Industrial compatible. The last four attributes are determined by the tax assessor's office. The assessed property value computed annually by the assessor's office and is used to determine the tax owed. Market readiness assesses the development availability of industrial zoned lands. The market readiness is defined as "Short Term" if the property is currently zoned for an industrial type use (e.g. Heavy Industrial, Light Industrial), with the development review process theoretically beginning within 2 months of a submission of an application by a firm. A "Long Term" status is given to properties with zoning currently incompatible with an industrial use, and therefore, the parcel would have to go through a rezoning process before the development review process. This entire process would take longer than 2 months. Two months is the minimum length of time it takes any development to go through the rezoning process.

The development potential gives us an idea on whether an industrial property is good for redevelopment or "brownfield" development. Development potential is determined by calculating the ratio of building value to land value and multiplying by 100. If the building value is 25% or less the land value, the property is deemed underutilized, therefore indicates a potential for redevelopment. Greater than 25% and less than 100% is deemed evaluated, which indicates a closer evaluation is needed to determine potential redevelopment. A ratio greater than 100% is defined as developed, indicating no real potential for redevelopment. A $0 building assessed value

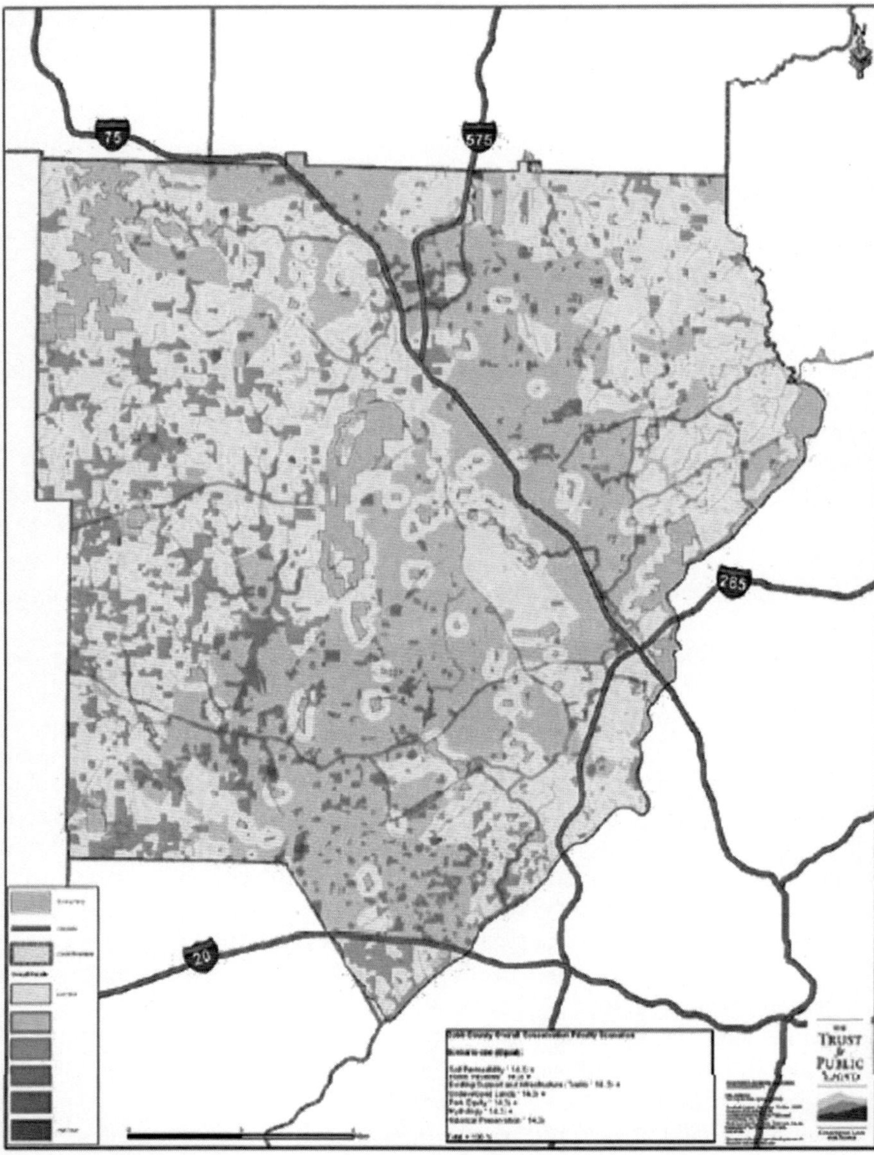

Fig. 13.3 Green space preservation priority map

is classified as undeveloped or vacant land. In all, there are 1,342 parcels classified as developed, 475 as undeveloped, 162 as evaluated and 376 as underutilized.

Spatial queries are executed by the model to determine distance to transportation networks and nodes, area of parcels, and distance to watercourses, parks, and residential areas. One, two and five miles buffers were created for freeways, major arterial roads, airports, bus stops and railways. Spatial queries then identified the parcels that fell into each buffer. Parcels located within 1 mile buffers for each transportation variable scored higher than those with farther out. For area of parcels, the minimum parcel size to be considered in the industrial assessment model is 50 acres.

Area-specific incentives are enterprise zones. An enterprise zone is an area designated by a local government as ready for development or redevelopment. Qualifying businesses locating or expanding within the zone are eligible for certain tax incentives and other economic development incentives for job creation and capital investment. Using GIS, these areas were determined by overlaying the enterprise zones over the industrial parcels and utilizing the "have their centroid in" feature of the "Select by Location" tool. While some industrial parcels were partially covered by enterprise zones, only those parcels whose centroids were within an enterprise zone were included.

Environmental constraints included floodplains, slope (greater than 25%) and wetlands. Each parcel zoned industrial was overlaid on these layers. If the parcel was located within one or more of these environmental constraint areas, the property was considered unsuitable for further development. Of the 2,537 parcels zoned for industry, 1,801 parcels (71%) did not have any environmental constraints. Once all of these factors were computed by the model and the various layers were overlaid, a total of 25 parcels were identified for future industrial development within the county.

Clearly the compilation of the FLUM would be very difficult if not impossible without the use of GIS. The FLUM is the culmination of many spatial and non-spatial data layers and data analyses. The spatial data layers used in the map range from the county boundary to parcels. Non-spatial data include zoning designation to assessed parcel value. The different types of applications discussed above serve to illustrate the scope and range of analyses conducted using GIS. From simple buffering to sophisticated modeling, using GIS was critical to assemble the FLUM. Moreover, as the FLUM is a dynamic document, future edits and additions will be easier to undertake with GIS.

13.7 Conclusion

In this chapter, we have provided an overview of major trends in the use of GIS by local governments. By and large, GIS has become more important and common place in local government. How GIS is structured and used within local government may differ tremendously; however, the trend toward server environments, such as interactive web base mapping, is evident, especially when resources, infrastructure, and personnel allow for it. In addition, the trend in data sharing has become more accessible and available both within the government structure (e.g. multi-agency, cross-departmental and statewide data sharing agreements) as well as to the public at large (e.g. Internet).

By looking at the government of Cobb County, Georgia, we see the modified decentralized organizational structure allows for government agencies or departments to assume ownership/responsibility for collecting, maintaining and updating GIS data. The county also uses the internet to disseminate some spatial data to the public at no cost. The public can also purchase certain GIS datasets online. This is part of the county's cost recovery program, which is common with many other local government GIS departments.

The power of GIS is its ability to visually display the comprehensive plan. In doing so, the vision, mission and goals become clearer, which leads to a better understanding by all parties of the current situation and what these parties would like to see in the future. As the comprehensive plan is dynamic, GIS can facilitate edits and modifications to the maps to reflect changing needs and desires of the county. Evaluating comprehensive plans, such as Cobb County, are ideal in that they are legally binding, spatially oriented, and are the framework to understanding the economic and environmental landscape under a local government jurisdiction (Brody et al. 2006).

Comprehensive planning is one of the most wide ranging activities undertaken by local governments. It involves several county agencies and requires public input. The maps created using GIS serve as communication devices to facilitate discussion at public meetings. Indeed, these maps were vital in the Public Participation Program of the comprehensive plan. Using Kelly and Becker's (2000) three criteria for a successful comprehensive plan, it is apparent Cobb County's plan is geographically comprehensive, as not only does it cover issues within the county, but is in line with larger regional planning concerns. It addresses planning issues at scales ranging from the county to street intersections. The plan also is long range in scope, as the FLUM is projected to 2030. As we argued in the introduction for a fourth criterion, we believe it is important for the county government to disseminate information before, during and after the creation of the comprehensive plan. The county has used the Internet to provide information to the public at these three time periods, including making available the data layers necessary to create the FLUM. The use of the Internet for disseminating information and data among departments and providing an online mapping application helps define Cobb County as an "enterprise-wide system" (Tomlinson 2003).

In the comprehensive plan document, the Future Land Use Map serves as the official future development map for the county, which reflects the desires and goals of the public and elected officials. The map is used to help guide development in the county. Any development proposals which are counter to the FLUM will find approval more difficult by zoning board. However, as noted earlier the FLUM is a living document and subject to modification. In fact, the comprehensive plan states, "there are circumstances when decisions will be made that are contrary to this document based upon a change in market conditions, information unbeknownst to staff..." (Community Development Agency 2006, p. 18).

The FLUM itself is a map comprising of a series of data layers, each representing a certain theme. Such themes include land use zoning, transportation networks, and hydrology, to name but a few. The benefit of using GIS to create the FLUM lies in part in the ability to create and apply models to the spatial and non-spatial data to answer questions. These questions may be simple in scope such as whether a proposed development lies within a wetland. Questions may be more complex as well, such as identifying parcels for potential industrial development based on a set of criteria. We have provided examples of the use of GIS to answer such questions in the compilation of the FLUM and ultimately the comprehensive plan. For example the industrial assessment model used by the county found 25 parcels zoned industrial or industrial compatible as parcel suitable for future development. These 25 parcels represent less than one percent of the industrial parcels in the county. Without using GIS identifying these parcels would have been time consuming.

Finally, it should be noted that although the significance of the FLUM is clearly stated in the comprehensive plan document, the mention of the use of GIS to compile, manage, manipulate and display data is almost non-existent. The first mention of GIS in the comprehensive plan is on page 148, in which it notes that paper records are to be converted into digital format for GIS usage. The near lack of mention of GIS may suggest just how ubiquitous the use of GIS is in Cobb County's government. The county's GIS department estimates there are nearly 160 county employees using GIS and that the county's internet map server creates over 100,000 maps a month for the public and 50,000 maps for county employees (Fail and Scharff 2008). There is no doubt that GIS in the county government is important for a wide variety of tasks ranging from daily crime mapping to comprehensive planning. Each task would be difficult to do without GIS. The county's GIS department recently updated its vision statement to read: Make GIS a transparent technology that is used routinely to create, manage and analyze data. We believe the GIS department has successfully achieved this.

Acknowledgments We would like to thank Phillip Westbrook, Tim Scharff and Dana Johnson from the government of Cobb County.

References

Bresnahan P (2005) Emerging issues, trends and technologies. In: Fleming C (ed) The GIS guide for local government officials, ERSI Press. Redlands, CA

Brody SD (2003) Implementing the principles of ecosystem management through local land use planning. Population and Environment 24:511–540

Brody SD, Highfield WE, Thornton S (2006) Planning at the urban fringe: an examination of the factors influencing nonconforming development patterns in southern Florida. Environment and Planning B: Planning and Design 33:75–96

Budić ZD (1994) Effectiveness of geographic information systems in local planning. American Planning Association 60:244–263

Budić ZD, Godschalk DR (1994) Implementation and management effectiveness in adoption of GIS technology in local governments. Computational Environment and Urban Systems 18:285–304

Ceccato V, Snickars AF (2000) Adapting GIS technology to the needs of local planning. Environment and Planning B: Planning and Design 27:923–937

Chan TO, Williamson IP (1999) The different identities of GIS and GIS diffusion. International Journal of Geographical Information Systems 13:267–281
Community Development Agency (2006) Mapping our future: 2030 comprehensive plan. Cobb County, Georgia.
Croswell PL (1994) A new GIS generation dawns in state government. GIS World 7:40–41
Dueker K, DeLacy PB (1990) GIS in the land development planning process balancing the needs of land use planers and real estate developers. Journal of the American Planning Association 56:486–492
Drummond W, French, SP (2008) The Future of GIS in Planning: Converging Technologies and Diverging Interests. Journal of the American Planning Association 74:161–174
Fail C, Scharff T (2008) Cobb County GIS: beyond the map. Southeastern Region Users Group Conference 2008. Jacksonville, FL
Flax L, Jackson R, Stein D (2002) Community vulnerability assessment tool methodology. Natural Hazards Review 3:163–176
Kelly ED, Becker B (2000) Community planning: an introduction to the comprehensive plan. Island Press, Washington, DC
Liu S, Zhu X (2004) Accessibility analyst: an integrated GIS tool for accessibility analysis in urban transportation planning. Environment and Planning B: Planning and Design 31:105–124
Nedović-Budić Z (2000) Geographic information science implications for urban and regional planning. URISA Journal 12:81–93
Nedović-Budić Z, Pinto JK (1999a) Interorganizational GIS: issues and prospects. The Annals of Regional Science 33:183–195
Nedović-Budić Z, Pinto JK (1999b) Understanding interorganizational GIS activities: a conceptual framework. URISA Journal 11:53–64
Nedović-Budić Z, Pinto JK, Warnecke L (2004) GIS database development and exchange: interaction mechanisms and motivations. Journal of Urban and Regional Information Systems Association 16:15–29
O'Looney J, (2000) Beyond maps: GIS and decision making in local government. ESRI Press, Redlands, CA
Petit C, Pullar D (1999) An integrated planning tool based upon multiple criteria evaluation of spatial information, computers. Environment and Urban Systems 23:1–19
Rogers EM (1983) Diffusion of innovation. 3rd edn. Free Press, New York
Sandström UG, Angelstam P, Khakee A (2006) Urban comprehensive planning – identifying barriers for the maintenance of functional habitat networks. Landscape and Urban Planning 75:43–57
Sarkar D (2003) Local governments use GIS. Federal Computer Week May 16
Waite B (2005) Planning, implementing and funding a GIS. In: Fleming C (ed) The GIS guide for local government officials, ERSI Press, Redlands, CA
Warnecke L, Beattie J, Kollin C, Lyday W, French S (1998) Geographic information technology in cities and counties: a nationwide assessment. American Forests, Washington, DC
Zaltman G, Duncan R, Holbeck J (1973) Innovations and organizations. John Wiley and Sons, New York

Index